Amateur Radio

Your Guide to Start Leaning Everything About Ham Radio

(Acing the Amateur Radio Technician Class Test With Ease)

Harold Fortune

Published By **Phil Dawson**

Harold Fortune

All Rights Reserved

Amateur Radio: Your Guide to Start Leaning Everything About Ham Radio (Acing the Amateur Radio Technician Class Test With Ease)

ISBN 978-1-7750277-0-6

No part of this guidebook shall be reproduced in any form without permission in writing from the publisher except in the case of brief quotations embodied in critical articles or reviews.

Legal & Disclaimer

The information contained in this book is not designed to replace or take the place of any form of medicine or professional medical advice. The information in this book has been provided for educational & entertainment purposes only.

The information contained in this book has been compiled from sources deemed reliable, and it is accurate to the best of the Author's knowledge; however, the Author cannot guarantee its accuracy and validity and cannot be held liable for any errors or omissions. Changes are periodically made to this book. You must consult your doctor or get professional medical advice before using any of the suggested remedies, techniques, or information in this book.

Upon using the information contained in this book, you agree to hold harmless the Author from and against any damages, costs, and expenses, including any legal fees potentially resulting from the application of any of the information provided by this guide. This disclaimer applies to any damages or injury caused by the use and application, whether directly or indirectly, of any advice or information presented, whether for breach of contract, tort, negligence, personal injury, criminal intent, or under any other cause of action.

You agree to accept all risks of using the information presented inside this book. You need to consult a professional medical practitioner in order to ensure you are both able and healthy enough to participate in this program.

Table Of Contents

Chapter 1: Commission's Rules E1a Operating Standards 1

Chapter 2: Miscellaneous Rules 18

Chapter 3: The Choices Listed Are Accurate 52

Chapter 4: Coordinate Systems 76

Chapter 5: Optical Components 100

Chapter 6: What Is Ham Radio? 113

Chapter 7: Ham Radio Made Simple 132

Chapter 8: Key Concepts 144

Chapter 9: Advanced Concepts 158

Chapter 10: Your First Contact 170

Chapter 1: Commission's Rules E1a
Operating Standards

The signal you receive is bigger than you imagine, and you must be cautious not to go too close to edge of the band. This is the point of an array of issues.

If you're using LSB the signal you are transmitting is at or below the frequency displayed. If you're operating LSB AFSK in the 17-meter band, which is a data segment that ranges from 18.068 up to 18.110 MHz, it will be unlawful to transmit 18.068 MHz. The LSB signal will be extended below the lowest frequency. This question will give an answer.

A SSB phone signal is approximately 3 kHz in width. The frequency of the carrier must be at least 3 kHz higher than the edge of the band below if you are using LSB. The LSB signal will extend 3 kHz less than the carrier.

The highest legal frequency for carriers in the 20-meter band for USB digital signals with one kHz bandwidth will mean 14.149 MHz. The answer is too complex. Keep in mind that the top edge of the sub-band digital is 14.150. If your USB signal measures one kHz in width, then it must be lower than 14.149.

If you are able to hear a radio station making a CQ in 3.601 Mhz LSB It is not legally permitted to answer the call since your sideband will be extended beyond the edge of the band. The limit of voice band is 3.6 milliseconds. The LSB signal will extend beyond the threshold of 3.6 MHz.

The 60-meter band is shared with other providers, which means the rules are a bit different. The maximum amount of power allowed on 60 meters can be 100 watts in effective radiation power, as opposed to a dipole half-wave. Be aware of 100 watts and dipoles on 60 meters.

At 60 meters, you can choose the CW carrier frequency to your center frequency for the channel. The 60-meter band is the only one that allows amateurs to use channels.

The highest power allowed on the 2200-meter band is one Watt EIRP (Equivalent isotropic radiated energy).

The power limit for the 630-meter band will be 5 Watts EIRP.

The use of phones is permitted throughout the entire band of 630 meters.

Stations that forward messages (like Packet) are usually controlled by automated. When a station sends out messages that are in violation of FCC rules, the controlling person of the original station is responsible for the infraction.

If your digital message forwarding system is able to accidentally forward a communication which was in violation of

FCC rules, it's best to end the forwarding in the earliest time possible.

Slow scan TV can be restricted to segments of the phone band. Slow scans are about 3 kHz, and is similar to the SSB telephone signal.

You are able to work in an airplane or on a vessel However, you have to get permission from the owner of the ship or the pilot who is in charge of the plane. The permission must be granted by the supervisor.

For an US licensed vessel operating in International waters, any FCC issued amateur license is acceptable. No endorsements are required.

Anyone who has an FCC-issued amateur radio license, or has been granted permission for foreign reciprocal operations can be under the control of a physical station located on a ship or craft that is registered in the United States. Make sure you look to find "FCC-issued amateur radio license" in both of the answers."

E1B Station Restrictions

The term "spurious" refers to any emissions that is not within its bandwidth, which can be reduced or completely eliminated, without altering the content of information sent. It is a sign that it's outside the bandwidth that is required for it.

An acceptable frequency to use Digital Radio Mondiale (DRM) that is based on voice is 3 kHz. Digital Radio Mondiale is digital audio. It ought to have about similar in width to SSB.

You have to shield the FCC monitoring center from damaging interference, if it is within a mile.

Prior to establishing an amateur base within a wilderness preserve or wildland area or an area included in the National Register of

Historic Places You must file the FCC with an Environmental Evaluation to the FCC.

The National Radio Quiet Zone is an area that surrounds the National Radio Astronomy Observatory. Trick: "National is in both the answer and question.

If you are planning to install an antenna on an area near an airport that is used by the public You may need be notified by that the Federal Aviation Administration and register it at the FCC.

PRB-1 covers zones in the state as well as local.

Local and state governments have to "reasonably accommodate" amateur radio. There are no such limitations for homeowners' associations. These restrictions can be excessive.

If the broadcast signal from an amateur station creates interference for broadcasters in domestic reception The FCC might impose limits on transmitting times on the frequencies that create the disturbance. The FCC could make "quiet times."

RACES stands for the Radio Amateur Civil Emergency Service which is a part of a protocol developed by FEMA as well as the FCC. Any amateur station licensed by the FCC and accredited by the civil defense group in the region served can operate under the RACES regulations. The RACES certification process is a good idea.

The frequencies that are authorized by RACES for amateur radio stations RACES rules include the frequencies for amateur service that are authorized to the controller. It is not a special frequency.

If a repeater is causing interference to a radiolocation system the operator of control must stop operating or modify the system to minimize the interference. This is why you shouldn't be able to disrupt the radio's location system.

E1C Control

The bandwidth of data transmissions over 60 meter is 2.8 KHz. Similar to SSB voice.

Communications that are not directly related to the mission of the service for amateurs as well as personal comments of a kind are examples of communications that can be sent to amateur radio stations from other countries. Same as domestic.

A IARP can be described as an International Amateur Radio Permit that lets US amateur radio operators to use their equipment within certain countries in the Americas. This is a multi-country licence.

Control operator duties for automatic control are distinct in comparison to local control since the automatic control system does not require the operator of the control isn't needed to be present at the point of control.

The station controlled by an automated system could never initiate third-party communications.

The duration that can be allowed for the transmissions of a remote controlled station when its control link fails is three minutes. It is required to stop transmission within 3 minutes.

CEPT is a convention that allows US amateurs to work within European countries and vice versa.

To be able to work within CEPT rules, it is necessary to be able to bring along a copy FCC Public Notice No. 16-1048. The document lists nations and the rules in various languages. Remember to read "FCC Public Notice."

The bandwidth of a signal is measured at 26 dB lower than its median power threshold.

The modulation index that is the highest allowed for angle modulation that is less than 29 MHz is 1. Angle modulation can be described as FM. FM Modulation Index. FM Modulation index is equal to the ratio of frequency shift in relation to the frequency of modulation. If the ratio is 1, it's very narrow.

The permissible mean power of an emission with a spurious component is 43 decibels below that of the basic emission. This is too difficult. Be aware that emissions from spurious sources must be less than 43 decibels.

Telephone emissions are allowed in the whole 630-meter band.

Power-line communications (PLC) is a method of transmitting data through electricity lines for distribution of power to customers. The operation of 630 or 2200 meters may cause problems.

If you plan to operate in the 2200 or 630-meter band You must inform your local Utilities Technologies Council of your number and the coordinates of your location. If you've have never heard of the Utilities Technologies Council before, that's the right choice.

It is required to wait for 30 days following notification to UTC to be able to transmit in the event that you haven't been informed that your station is within 1km of the PLC systems that use the same frequency. It's too complicated. Find the solution with "wait 30 days."

E1D Satellites

Telemetry is the transmission in one direction of measurement. The key word is "metering".

Telecommand signals emanating from a space-based telecommander could transmit specific codes that obscure the purpose of messages. The pulses and beeps don't convey any meaning to the listener.

Space telecommand stations are an instrument that sends out communications to start, alter and/or end the function of the space station. In other words, it commands the space station to perform some thing.

A telemetry station that is balloon-borne must be identified with a number. Like everyone else.

The station that is operated through the telecommand system within 50km of earth's surface should post the station's location:

Photocopies of the license issued by the station

Labels with the address, name and phone number of the licensee for the station.

The label should contain the name, address and phone number for the operator of the control.

* All of these options are true.

This information will help in locating the owner.

The maximum amount of power that can be used in the operation of the model craft via Telecommand is one watt.

The bands that are HF-licensed for space stations include 40 millimeters, 20 millimeters 17 m, 15 12, as well as 10 m bands.

Tips: All of 40-10 with 30 m.

On VHF just 2 meters of spectrum is accessible to space stations.

On UHF, 70 cm as well as 13 cm can be used for space stations.

Every amateur station that is approved by the licensee of the space station can be considered Telecommand stations. It is important to get approval from your boss, the person who licenses space station.

Stations that are eligible to be operated as Earth stations include amateur stations, but dependent on the authority of the operator.

The stations that be transmitting one-way communications include a beacon station, space station, or telecommand station. The key is to find an answer using telecommand or one-way control signals.

E1E Examiners

Consider yourself lucky that you don't have to go to the FCC Field Office to take the exam. The FCC transferred the responsibility of testing to volunteer Examiner Coordinators (VECs) in the year 1984. There are fourteen Volunteer Examiner Coordinator organizations, who jointly develop the pools of questions and run the test system. On the day of testing people you encounter are volunteer Examiners (VEs) who are accredited by the coordinators (VECs).

The VEs can be reimbursed for the costs of preparation or processing, conducting, and coordinating examinations. They are not reimbursed for training or teaching.

Questions for the examinations written U.S. amateur license examinations are included in a pool of exam questions that are maintained by the VECs. Examiner Coordinator organizations manage the pools. Examiner Coordinators manage these pools, not examiners themselves.

The Examiner Coordinator is a volunteer Examiner Coordinator is a group who has signed an agreement with FCC in order to organize, organize and oversee exams for amateur radio licenses. Hint: A coordinator coordinates.

It is the Volunteer Examiner accreditation process is the process through which an VEC ensures that a VE applicants meet FCC standards. The key to remember is that a VE is certified through an VEC but not by the FCC.

The minimum score for passing the tests is 74%..

The person administering the VE is responsible for correct conduct and the necessary surveillance during a radio amateur exam session for licenses. Each person is responsible to ensure the reliability of the process. We do not want the FCC to conduct examinations inside the agency.

If a test taker fails to adhere to the exam's guidelines then the examiner must immediately end the test.

A VE is not allowed to administer an examination to anyone that is related to the VE as defined in FCC regulations. The rule is that employees and friends are fine.

A VE who knowingly conducts an exam or certifies it is liable to lose his amateur station license as well as his amateur operator's license. The penalty is not jail time or fines The loss of license can be even more serious!

Following the successful completion of the exam after completing the exam, VEs have to submit their application for approval to the coordinator VEC in accordance with the guidelines of the VEC. The exam is graded in-person by VEs and they then send the result for their VEC. The VEC is able to load the results into FCC database. FCC database.

In the event that an examinee receives an acceptable grade, three VEs are required to confirm that the applicant has the qualifications for a license grant, and also that they have fulfilled the VE requirements. In the search results, look at "three VEs," and you will find the solution.

Chapter 2: Miscellaneous Rules

Spread-spectrum signals change the frequency of transmission in a specific design to minimize the amount of interference, and to ensure security. This results in a broad spectrum, and therefore it's restricted to larger and higher spectrums. Spread-spectrum signals are only permitted at frequencies for amateurs that are higher than 22 millimeters.

A Canadian licence holder has similar privileges as those in the US without exceeding what is allowed under the US Amateur Extra rights.

Amplifiers need to be FCC certified to safeguard from CBers' misuse as well as to ensure purity of the spectrum. The dealer can sell an RF power amplifier externally that operates lower than 144 MHz, but does not have FCC approval if the amplifier was bought in a used state by an amateur operator, and then sold to an amateur operator, to use on the station of that operator. It's too complicated! Just remember, amateur to amateur is OK.

Line A runs that is roughly parallel to the south and north of the border between Canada and the United States.

Amateur radio stations within the US are not permitted to broadcast on 420 MHz or the 430 MHz band if they're located north from Line A.

A Temporary Special Authority could be granted through the FCC to permit the development of experimental communications between amateurs. They typically allow use outside the normal frequency of amateur radio.

A broadcaster from an amateur station can transmit messages to businesses but only if neither the station or his employer have any financial interest in the message.

Communication that is made to be used for hire or as compensation, unless otherwise stipulated in the regulations, are not permitted. The key word is "For hire" is prohibited. Hams aren't paid as Ham. Ham.

Amateur radio shouldn't be used for commercial purposes. For example, I'm traveling to the house of my friend Bob to assist him with his antenna. I phone him up on the radio and ask him if you would like to take pizza to his house. He replies, yes any kind you want. A different ham who owns the pizza shop, is able to hear me and rings. "Hey Buck, what kind of pizza would you prefer. I'll be able to make it to go." It's not a good idea The same pizza, but an entirely different outcome.

If you intend to make use of spread spectrum these conditions are applicable:

* Should not create harmful interference

* Must reside in an area that is regulated through the FCC or a nation that allows SS emission

* Cannot be used to obscure intent

* All choices are true.

An auxiliary station can be described as an entity that is controlled by another station through an internet link. The operator who is in charge of the additional station can include a Technician, General Advanced, a Technician, or an Amateur Extra operator. Advanced is an older license class. Find the answer that mentions "Technician."

To be eligible for the approval for FCC certification, the amplifier must meet the FCC's bogus emission standards when operating at 1500 watts, or its maximum output power

Tips: In order to obtain approval, meet the FCC. Do not bother to memorize the information.

E2 - OPERATING PROCEDURES

E2A Amateur Radio in Space

The ascending path of an amateur satellite runs between the south and north. The trick is to say: "Ascending." Think of it as a climb out of towards the South Pole to the North.

A transponder is a device that receives signals and emits an electromagnetic signal in reaction. The satellite could detect a signal in the one frequency (uplink) and retransmit the signal on another band (downlink).

In the case of a satellite using the inverting transponder

The Doppler shift has been reduced due to the fact that the downlink and uplink shifts occur in opposite directions.

* Signal location within the band has been reversed

* The upper sideband of the uplink is replaced by lower sideband in the downlink and reverse.

All of these options are accurate.

Note: Everything that is written in the answer sheet is reversed.

The signal is flipped by the inverting transponder since the signal goes through a mixer. it is the difference, not the whole is transmitted. Hint: Inverted.

Satellites operate in that of the downlink and uplink frequency bands.

The letters on the satellite's mode designer indicate the frequency of uplink and downlink.

If a satellite operates in U/V mode indicates that the uplink is UHF and the downlink will be VHF.

Keplerian elements are parameters used to describe the orbit of satellites. Hint: Kepler developed formulas for describing an orbiting body.

The linear transponder is able to be used to relay:

* FM and CW

* SSB and SSTV

* PSK and Packet

The choices listed are accurate

It is a linear device, which means it can transmit any signal.

The power radiated to the satellite using a linear transponder must be restricted to prevent reducing the power of downlink to other users. If a satellite receives a powerful signal will have a good connection. It will also reduce the power it transmits to save energy.

The words "L band" and "S band" are used to describe the bands of 23 centimeters and 13 centimeters. Make a note of the numbers 23 and 13, and you'll know the answer.

Satellites that remain within a specific position in the sky is called geostationary. It is important to note that if it remains only in one spot then it's stationary.

In order to minimize the effects of the spin modulation process and Faraday rotation, choose the circularly polarized antenna. Trick: Spin around on circles.

Digital forward and store functions is to save digital information in satellites in order to later be downloaded by different stations.

Tips: The digital storage and forwarding store digital messages, and then forwards messages afterward.

Low Earth orbiting digital satellites send messages across the globe via storage and forward. The key word is "Around the world" requires storage of the message until they reach the desired region.

E2B Television

Fast-scan (NTSC) TV transmits 30 frames every second.

Fast-scan (NTSC) TV transmits 525 lines in each frame. Fast-scan comprises 30 frames that contain 525 lines in each frame.

A pattern of interlaced scanning is created in a fast-scan (NTSC) TV by scanning lines with odd numbers in one of the fields and even numbered lines in the following. The clue is that odd and even patterns are interlaced.

Information about colors on analog SSTV is transmitted via colors that are transmitted sequentiallyeach color being followed by another.

Vestigial sideband modulation can be described as an amplitude modulation where one whole sideband and a part of another are broadcast. It is the remnant of an area of the other sideband.

The advantages of having a vestigial sidebands used in standard speed-scan TV is that it minimizes bandwidth, while also allowing simple circuits for video detection.

Since only a small portion of the second sideband is sent which reduces capacity. Hint: "Reduces bandwidth."

The component of the signal that transmits colors is known as the chroma. The word chroma is used to describe either relating to or created by the use of color.

The image's brightness is encoded in an audio frequency.

The method that permits commercial TV receivers with analog input to be utilized for rapid-scan TV within the 70cm band broadcasts on channels that are that are shared by cable TV. Warning: too much data!

Commercial TV receivers are able to take in the cable television.

A radio with SSB capabilities and a compatible computer are required to decode SSTV by using Digital Radio

Mondiale (DRM) No additional equipment is required. DRM is interpreted by computers.

This Vertical Interval Signaling (VIS) code is used in an SSTV transmission in order to determine the SSTV mode used. The code is actually signaling the SSTV mode.

SSTV receiver software is alerted to initiate the new line using particular tone frequencies.

Be sure to look at "tone frequencies" in answers.

E2C Operating Methods; Contest and DX

In the event that you are a U.S. licensed operator is using a remote control transmitter within the U.S., no additional indication is needed (when it comes to identifying).

Self-spotting is often a prohibited method of putting your personal name and number on the spotting networks. There is no way to identify yourself when playing.

Contests of amateur types are usually not allowed from 30 meters. Contests are not permitted in all among the WARC bands (30 17 and 12 meters).

A mesh network connects by wireless signals using a adapter for networks. The frequency used in mesh networks can be shared among numerous wireless data networks that are unlicensed. The wireless network adapters are not licensed.

The most common equipment used for implementing the ham radio mesh network is a wireless router that runs custom software.

The method that individual nodes employ to create a mesh network is link discovery and protocols. It is important to note that a mesh network is a way to discover the other nodes in its network and creates hyperlinks

An DX QSL manager handles the receipt and the sending of confirmation cards to the DX station. [3]

This U.S. QSL bureau system can be used to establish contact between the U.S. station and a non-U.S. station. It is important to note that the U.S. QSL bureau system is not able to handle confirmations from the domestic market.

In a contest between VHF and UHF You should expect to observe the greatest amount of activity within the weak signal portion of the band. This is because it has the highest level of the activity occurring near the call frequency. The majority of people be looking for weak signals around the call frequency that is weak.

Cabrillo formats are the preferred format for electronic logs to be submitted. A computer program for logging can convert your log file into Cabrillo format and upload to the contest's sponsor.

A DX station could say they're listening to distinct frequency

* Since they transmit at a frequency that is not allowed to one or more of the stations that are responding.

* To distinguish between the callers and the DX station.

* To increase the operational effectiveness of the system by lessening interference.

These choices are accurate.

There are many DX stations work "split," listening on the same frequency as their broadcasts. Find my book on the subject, "How to Chase, Perform and Confirm DX. - The Easy Way."

In order to call the DX station in a competition or pileup, you'll normally identify the station by sending a entire call twice or once. No partial calls, no grid squares, no repetitive identifying.

E2D Operating Methods: VHF/UHF

The digital mode developed for meteor scattering is MSK144. Hint: Meteor SKatter

An effective method for making meteor scatter contacts is to:

* 15 second sequences of timed transmission that alternate stations depending on their location.

* Utilization in High-Speed CW as well as digital mode.

* Short message with quickly repeating call numbers and signals reports.

* All of the options are accurate.

A hint: Many good strategies, therefore all are accurate.

The way to establish EME contacts is to use time-synchronized transmitting alternately by each station. For contact to be established, it is necessary to determine the time to be listening (time time synchronized).

The most effective digital mode in EME communications can be found in JT65. EME is Earth-Moon and Earth, and as neither one moves fast it is possible to use slow modes, such as JT65. It is designed for very weak signal strength.

The modulation type that JT65 contacts use includes multi-tone AFSK. AFSK is Audio Frequency Shift Keying. Hint: Remember JT65 uses audio tones.

One benefit that comes with JT65 mode is that it has the ability to recognize signals with a small signal-to noise ratio. It is important to note that JT65 detects signals at a lower level than the noise.

The method used to monitor live, on a real-time basis, balloons containing amateur radio transmitters is known as APRS. APRS is Automatic Packet Reporting System. The GPS connects to your radio in order to transmit and receive information about your position.

The protocol that is used in APRS uses is AX.25.

The kind of frame that is used to send APRS beacon information includes Unnumbered Information. Information: The data are data.

A APRS station could help in supporting public service communications since the GPS station can transmit details to indicate the location of a mobile station during the occasion.

The data that is used by an APRS network to transmit your location is latitude and longitude. The latitude and longitude are the two ways to determine your current location.

E2E Operating Methods: HF digital

One of the most popular types of modulation used for data emission less than 30 MHz, is FSK. Frequency Shift Keying. FSK is utilized to create RTTY (teletype). FSK isn't the sole kind of data transmission mode however it's the only option that's accurate. Note: Be aware the fact that FSK can be described as a digital format.

The letters FEC refer to Forward Error Correction. Data that is redundant is transmitted, in the event that it does not match an existing request, the re-send procedure follows.

Its timing for FT4 contacts is controlled with alternating transmissions that occur in 7.5 seconds intervals. FT4 is a type of data that sends data for 7.5 seconds before listening at 7.5 seconds.

If an ellipse on an FSK displayed crossed-ellipse suddenly vanishes it is likely that selective fading occurred. Indication: The ellipse disappeared.

The mode digital that does not support keyboard-to-keyboard operations is called PACTOR. It is a one-way mode. The entire message can be transmitted uninterrupted. This isn't an online chatroom.

The most commonly used frequency of data for an HF packet is 300 baud.

Digital mode that has the most data throughput in good communication conditions is a 300 baud.

Tips: Search at "300 baud" in an answer.

If you're trying establish contact with a digital station that operates on a clear frequency, but you are not successful it could be because the following: * The frequency you are transmitting is not correct.

* The version of the protocol that you're using isn't compatible with the digital station.

A different station that you're not able to hear is the same frequency.

These choices are true.

There are many reasons why this might not be working. If you are able to identify two reasons, "all of the above" is your answer.

The digital mode used by HF for the transfer of binary data is called PACTOR. PACTOR transmits and receives digital information via radio.

PSK31 utilizes variable length coding to improve high bandwidth efficiency. There are different letters of various lengths. Capital letters are more than twice the time to write therefore, don't make use of the entire caps.

PSK31 is the narrowest bandwidth (31Hz)

The distinction in FSK that is direct FSK or audio FSK is that direct FSK transmits an information signal into the transmitting VFO whereas AFSK sends out tones through the phones. Variating the frequency (FSK) modifies the tone at the receiver part. Audio FSK sends the audio tone directly into the microphone. Then, it's transferred via SSB.

ALE stations establish contact through continually scanning for frequencies and activating the radio once the appropriate number is heard. ALE is Automatic Link Enable. The radio is scanned and then automatically connects to.

E3 - RADIO WAVE PROPAGATION

E3A Electromagnetic Waves

The maximum distance between two stations connected by EME will be approximately 12,000 miles in the event that the Moon is visible to both stations. Remember the miles. The only way for Earth-Moon-Earth to function is when the Moon can be seen by both stations. It is common sense.

A fading in the Libration of an EME signal is an blurry, irregular fade. It is the perception of oscillating movement of two bodies orbiting. The motion would appear irregular.

If you are planning for an EME connection, the principal condition for the lowest loss in path is at a time when the Moon is perigee. Perigee is the closest point to Earth which means that your signal is less distant. Apogee lies further.

Hepburn maps forecast the likelihood of propagation through the tropospheric sphere.

The propagation of the Tropospheric wave is usually seen in cold and warm fronts. A difference in temperature can bend radio waves.

If DX signals get less than adequate to duplicate within a couple of hours following sunset, change to a lower-frequency band. At night, the lower frequencies will come alive.

Atmospheric conduits capable of transmitting microwave signals typically form on lakes composed of water. It is important to note that microwaves bounce off the water's vapor.

Meteor scatter occurs due to free electrons that are present in the layer of E. The trick is to: Meteor, free, electrons = a lot of Es.

Meteor scatter is best suitable for frequencies between 28 and 28 MHz to 148 MHz. It is important to note that six meters are the most popular meteor scattering, and this is the only answer with the 50-MHz range (six meters). "Meteor" has six letters which aids me in remembering "six meters."

The structural structure in the air that is able to provide a pathway for the propagation of microwaves is called the temperature inversion. The signal is bounced between layers of various temperatures.

The most common range of tropospheric propagation ranges from 100 to 300 miles.

The Auroral Activity is triggered by interactions in the E layer of charged particles that come from the Sun in the earth's magnetic field. The key to understanding this is: Aurora is a result of charged particles that originate from the Sun. It's not important to know which layer.

The ideal mode of Aurora propagation is CW. The key is that CW offers a 10dB advantages over SSB and always is the most effective mode of propagation when compared with the other options.

Circularly polarized electromagnetic signals are electromagnetic waves that are characterized by a rotating electrical field. Hint: Circles rotate.

E3B Transequatorial Propagation

Transequatorial propagation takes place the movement of two mid-latitudes with a distance of approximately from north to south along the magnetic equator. It is important to note that if it's transequatorial, it has to be crossing the equator.

The most likely maximum range in transequatorial propagation would be 5 miles.

The ideal time to experience transequatorial propagation is in the afternoon or even in the evening. Tip: Let the sun shine the whole day to recharge these electrons.

The term "ordinary" and "extraordinary" waves are referring to waves that are independent generated in the ionosphere. They are circularly Polarized. The two types of waves are distinct.

Linearly polarized waveforms that break into extraordinary and ordinary become the elliptically-polarized.

Tips: Search to find "elliptically polarized" in answers regarding extraordinary waves.

The long-path propagation can be provided by 160-10 meters. The HF bands provide long path support - that is, going through the Earth all the way.

The frequency band that typically offers long-path propagation is 20 meters. It is important to note that the frequency typically available to DX is 20 meters.

The season when that sporadic E proliferating is likely to happen is during the solstices, particularly during the summer solstice. Sporadic E is a common summer phenomenon.

It is sporadic. E spread can take place at during any daytime. It's rare and may occur at anywhere at any moment.

The main aspect of the chordal hop propagation is that it has successive reflections of the ionospheric sphere without an intermediate reflection coming from the ground. The reflection of the signal inside the ionosphere, but does not reach with the Earth until it eventually is returned to the receiver.

The advantage of chordal hops is that the signal suffers lesser loss when compared with multi-hops that use Earth for a reflection. A tip: less loss is always a positive thing.

E3C Radio Propagation

Ray tracing is the process of modeling the radio waves' path across the Ionosphere. It is important to trace the path of the radio wave.

A increasing A index or K index is a sign of a growing perturbation of the geomagnetic fields. It is important to note that the index increases in response to geomagnetic disruptions.

If you notice that the A or K index rises it is likely that the pathways with the highest amounts of absorption are the polar. A higher A or K index is a sign of an electromagnetic disturbance in the geomagnetic field, it is most likely to be the strongest near the poles.

Bz is the direction and the strength of the magnetic field that is interplanetary. The two letters B and Z symbolize two things: both strength and direction. There are other solutions that only mean the same thing.

The direction of Bz which enhances the probability that particles coming from the Sun can cause disturbances is to the south. It is important to note that if the particles heading toward the south (to to the south) then they'll hit at the North magnetic pole, causing interruptions.

It was discovered during the Technician Test that VHF and UHF radio waves be bent slightly above the Horizon. Radio horizons can be larger than the geometric horizon by approximately 15 percent of its distance. Cheat: It's not a huge amount but this is a low solution.

Solar flares are classified according to an alphabet. The highest intensity of solar flares is class X. Hint: X as in "eXtreme."

Space weather is also given an initial word. The word G5 is a reference to an extreme geomagnetic thunderstorm. Hint: "G" as in "geomagnetic."

The magnitude of an X3 flare is 50% higher than that of an 2 flare. The reason for this is that 3 is 50% more intense than 2.

The solar parameter 304A measures the UV emission in 304 angstroms. They are related to solar flux. Cheat: 304 A as in "Angstroms."

If the frequency rises as the frequency increases, the distance at which ground waves propagation diminishes. The ground absorbs greater frequencies.

The polarization type that is most suitable to propagate ground waves is vertical. A hint: Radio towers for AM are horizontal.

The distance of the radio-path horizon is greater than the geometric horizon as a result of the downward-bending effect of fluctuations in density of the atmosphere. It is important to note that the air is bending the wave above the Horizon.

An abrupt increase in background radio noise could suggest that a solar flare may have been observed.

E4 - AMATEUR PRACTICES

E4A Test Equipment

The maximum frequency signal that is able to be precisely presented on an oscilloscope digital is restricted by the rate at which it samples the converter from analog to digital. Note: The frequency of the signal is determined by the frequency at which an oscilloscope is able to sample. A more frequent sampling frequency means greater frequency.

The spectrum analyzer displays RF amplified and frequency. This is because it analyzes the spectrum to show the intensity and frequency that the signals.

In order to display spurious signals as well as intermodulation distortions inside the SSB transmitter, you can use an analyzer for spectrum. Tips: If you want to display unwanted signals, you can use the spectrum analyzer to identify the source of these signals.

A probe that calibrates an oscilloscope is described as "compensating." Compensation of the probe of an oscilloscope is usually performed by using the square wave, and then adjusting the probe so that the horizontal lines appear as flat as is possible. It is important to note that a square wave display displays straight lines that are symmetrical.

A prescaler in the frequency counter splits the higher frequency signal, so that it can be displayed by a counter that is low frequency. it. Tips: It will scale the signal prior to being taken into account.

Affects of aliasing on the digital oscilloscope that is that is caused by making the base too slowly can result in a false, slow-moving low-frequency signal being displayed. The key is to forget why it occurs. AKA: Aliasing refers to distortion or fake display.

If you are using an oscilloscope probe, it is a good idea to ensure that the signal ground lead as small as is possible. Note: A shorter ground leads has less impact on the measurement circuit.

The benefit of having the antenna analyzer over the SWR bridge is that antenna analyzers aren't dependent on an external source of RF. The antenna analyzer produces its own radio frequency.

The device which measures SWR is called an antenna analyzer.

An antenna analyzer must be directly connected into the feedline. Antenna analyzers are equipped with an coax jack that plugs into the coax.

For the display of various digital signal states at once using the logic analyzer. The signals used in logic circuits include high/low, or off/on.

E4B Measurements

The reliability that a counter uses will be dependent on the base's accuracy. Tips: The base should be exact.

The significance of sensitivity on a voltmeter which is expressed as ohms per volt is an accurate reading on the voltmeter multiplied by the ohms/volt. It will reveal the impedance of the input. Hint: Algebra. Volts x ohms/volt = ohms. Ohms indicate impedance. It is a trick: The term "voltmeter" is in both the answer and question.

Subscripts in S parameters are the ports or ports from which measurements are conducted. Cheat: S as in "portS."[4]

The S parameter that is equivalent to forward gain is called S21. Cheat: We viewed ahead to the day we turn 21.

The S parameter which represents SWR or return loss is called S11. Cheat SWR: 1:1 SWR is ideal.

To calibrate an the RF vector network analyzer you will need short circuits, an open circuit, and 50-ohm loads. It is recommended to calibrate the device at the extremes of the two and also you use the "normal" load of 50-ohm coax.

An analyzer of vector networks will be able to determine:

* Input impedance

* Output impedance

* Reflection coefficient

Chapter 3: The Choices Listed Are Accurate

It analyzes the network's parameters of an electrical circuit. into, out and reflection.

The amount of power that is absorbs by the load, when it is forward powered by 100 watts, and the power reflected is 25 watts equals 75 watts. Hint: Simple math. 100 watts was emitted and then 25 returned. So 75 watts would be within the load.

The relative measure of Q in a circuit tuned to series is the frequency of the circuit's response to frequency. The higher "Q" means narrower bandwidth. The way to measure Q is by taking the bandwidth measured in the frequency. You can forget about "series tuned."

The reading of an RF ammeter connected to the line feeding the antenna is likely to increase when there's more power coming into the antenna. The key is to have more amps, which means greater energy.

To determine the intermodulation distortion (IMD) within the case of an SSB transmitter, you need to modulate using two audio frequencies that are not harmonically related and then observe the result in the spectrum analyzer. The trick is to modulate using audio and then measure the distortion by using the spectrum analyzer.

E4C Receiver Performance

The impact of excess frequency noises in the local oscillator within a receiver is that it may be combined with powerful signals at nearby frequencies, causing disturbance. The reason for this is that excessive noise can cause the interference.

The circuits in the receiver that are able to effectively eliminate interference from out-of-band signals can be a front-end filters or a pre-selector. Be aware that these signals that are not in the band need to be taken out from the front of the receiver before they reach the internals that are more sensitive to the receiver.

A roofing filter is installed prior to more sensitive circuits in the IF. A roofing filter with a narrow-band band affects the performance of receivers because it increases the dynamic range of the receiver by degrading significant signals in the vicinity of the receive frequency. Tips: Filters reduce signals close to the frequency of reception.

The name used for suppression of the FM receiver of a more powerful signal at the same frequency is called capture effect. It is important to note that FM receivers can are able to capture the highest signal.

The the noise figure of the receiver is the proportion in decibels of the noise produced by the receiver to its minimal noise that is theoretically achievable. The question is regarding the noise figure of the receiver. It is stated that the answer refers to noise produced through the receiver. In addition, the question provides the solution.

The amount of -174dBm/Hz noise floor is the theoretical noise level at the point of entry for a perfect receiver operating at the room temperature. Cheat is guaranteed to be the only option on the test that is able to include "room temperature."

A CW receiver equipped with AGC off feature has an output noise power of the equivalent of -174 DBm/Hz. The unmodulated level of carrier input for this receiver, which could result in an SNR for audio output of zero decibels for a noise bandwidth would be -148 dBm. The trick is to not be serious. Memorize -148 dBm

It is the MDS in a receiver represents the lowest discernible signal. It's an indicator of the sensitivity of a receiver.

SDR is a type of radio. SDR is a radio that has been programmed in software. A converter from analog to digital is at the core for an

SDR. It converts the signals into the digital format and also zeros. After that, computers sort these signals. A SDR receiver gets overloaded if signals are higher than the voltage at which the receiver is operating. digital-to-analog converter. Tips: Signals that exceed the voltage of reference are causing overload to the receiver.

Super-heterodyne receivers are the ones that transform the signal received into an intermediate frequency fixed (IF) and then feed it into filters specifically designed for this frequency. One reason to choose higher frequencies for the IF is that it's easier to the front-end circuitry to block image-related responses. Combining "up" to a higher frequency will cause images to be further from the intermediate frequency.

Mixing may produce images. If the frequency of intermediate is 455 kHz then the converter is 455 kHz either above or below the frequency that is tuned. It can also mix with other signals, creating images on the frequency that is tuned. An instrument tuned to 14.300 MHz and the frequency IF of 455 kHz will also be receiving signals at 15.210 MHz. Local oscillators will be operating at 14.755 Mhz to create an IF frequency of 455 kHz. Another signal, that is 15.210 will also blend with 14.755 and result in a 455 kHz output. The easiest way to do this is to increase the IF before adding it to the frequency that is tuned.

Reciprocal Mixing is the locally oscillator-phase noise that mixes with nearby strong signals for the purpose of creating the illusion of interference. The local oscillator phase noise triggers reciprocal mixing.

The benefit in having multiple the IF bandwidths for receivers is that the bandwidth of receivers is adjustable to correspond with the modulation frequency. It is possible to use an IF filter that is narrow to use for CW or a more broad one to use for SSB.

A attenuator is a device to decrease receiver overload, without a significant impact on the ratio signal-to-noise in the event that atmospheric noise is higher than the internal generated noise. Reduce noise only to what is generated by the receiver. This ensures that it is clearer.

The most significant factor that affects the SDR the dynamic range of a receiver is the digital-to-analog converter's sample size. Make sure to look for an digital-to-analog converters in your answer. A SDR requires one.

E4D Receiver Performance

The block dynamic range of an receiver is the amount of difference in decibels between that noise floor as well as the volume of an input signal that can cause one dB of gain compression. Note: How much of an incoming signal is it to increase the gain by one dB. Answer: Recall your answer using 1 dB.

Two issues caused by the low dynamic range of the receiver is unwanted signals caused by cross modulation of the targeted signal as well as desensitization due to neighboring signals. Tips: Strong signals adjacent to one another can cause a receiver to be unable to recognize them due to poor dynamic range.

Intermodulation interfering can happen in two repeaters when two repeaters are within close proximity and when the signals combine within the amplifier final of either or both. The key is to know that mixing occurs.

To minimize or completely remove intermodulation interference from the repeater, you should use an appropriately terminated circulator on the input of your transmitter. Hint: To eliminate, terminate. Circulator is a single-way valve which shunts out signals from the transmission line before sending the signals to a dummy load (properly ended).

If a receiver is tuned to 146.70 MHz and a nearby station transmits on 146.520 MHz, the transmitter frequencies which would produce an intermodulation-product signal in the receiver are 146.34 MHz and 146.61 MHz. Tip: Calculate the sum of these two (146.70 + 146.52)/2 then look for an solution that contains 146.61. There is only one solution however that's enough for you to select the correct solution.

A term used to describe spurious signals caused by the fusion of more than two signaling elements in a nonlinear circuit or device is called intermodulation. The reason for this is that intermodulation can cause spurious signals.

Electronic circuits are prone to intermodulation. It is result of non-linear circuits, or other devices.

The word used for a decrease in the sensitivity of receivers caused by strong signals near the reception frequency is known as desensitization.

One way to lessen desensitization is by reducing the bandwidth of the radio receiver. Tips: Use the filter with a smaller size to cut the signal that causes desensitization.

The function of a preselector within the receiver is to improve the rejection of signals that are not in the band you want to exclude. The key is to preselect signal.

Third-order intercept levels at 40 dBm implies that the pair of signals that are 40 dBm produce a third-order intermodulation product that has similar amplitude to in the signals input. It is a cheat: Two signals of the same signal.

Intermodulation with odd-order order is very interesting as the odd-order products from two signals within a specific band is also most likely to be in the range.

It is possible that you are attracted since the product belong to the band which you're working with.

E4E Noise Suppression and Grounding

The drawback of notch-filtering automatically for CW signals is that it eliminates the CW signal while simultaneously. It is important to note that the automatic filter eliminates all tones, even the tone CW.

A way to reduce noise through a digital signal processor noise filtering includes:

* White noise from Broadband

* Ignition noise

* Power line noise

These choices are true.

Tips: Digital signal processing can be effective against a variety of background noisy.

A receiver noise blanker may have the ability to block signals that are visible across an extensive band. The noise blanker can be efficient against electrical fences as well as spark plugs.

The radiated and conducted noise of an alternator for cars is easily squelched by connecting the power lead of your radio straight to the battery and installing coaxial capacitors along the lead of the alternator. It is recommended to connect direct to the battery can be the most efficient method to power your mobile device. Find that information in the question.

The noise of an electric motor is able to be reduced by putting the AC-line filter with brute force connection with the motor's leads. The trick is to suppress noise by using brutal force.

The proximity of a computer could cause irregular modulated or unmodulated signal at particular frequency. A hint: Computer circuits may create "birdies" - signals heard at certain frequency.

Shielded cables may emit or experience interference from common mode currents in the conductors as well as the shield. This can be common to the conductors and shields. Solution: Put an insulator made of ferrite on the cable.

A current that flows in a uniform manner on every conductor is known as common-mode current. The key is that it's common for all conductors.

A negative effect of employing an IF noise-blocker is that the signals can appear broad even though they comply with the emission requirements. Trick: The IFR (intermediate frequency) operates at radio frequency. Other answers are all related to audio. If you're unable to recognize the right answer, you need to read the entire question and answer carefully to find clues.

A loud buzzing or roaring noise which comes and goes can be:

* Arcing contacts inside the thermostatically controlled device.

* Defective doorbell or doorbell transformer.

* Illuminated display malfunctioning.

These choices are true.

Tips: They all run intermittently and so noise is heard and then gone.

If you can hear a mix of AM stations from your local area that are in the HF and MF bands, it originates from close-by corroded metal joints mixing broadcast signals. Metal joints that are corroded can function like diodes or mix signals.

E5 - ELECTRICAL PRINCIPLES

E5A Resonance and Q

The voltage across the reactance of a the series could be higher than the voltage that is applied due to resonance. These high voltages could create arcs in the antenna tuner. Resonance is thought of as the process of multiplying.

Resonance of an electrical circuit refers to that frequency when capacitive reactance is greater than the inductive reaction. If the two reactances are identical both are equal, they cancel each other which causes the circuit to resonate.

The impedance of a series RLC circuit in resonance is about equivalent to the resistance of the circuit. The key is: "RLC" refers to the inductor, resistor, and capacitor. If inductor and capacitor do not cancel each other the only thing left is the resistance.

The amount of impedance produced by the resistance, inductor and capacitor, all parallel at resonance is the same as the resistance of the circuit. The answer is: whether in series or parallel, when you are at resonance, it's exactly the same. The only thing left to do is resistance.

"Q" stands for reactive Q, which is a measurement of quality of. High Q circuits are highly selective and narrow banded. What happens when you increase the Q of an impedance-matching network is that the bandwidth for matching will decrease.

The circulation current of an LC circuit that is parallel LC circuit during resonance is the highest. In other words, the current circulating in the circuit is highest. Inductive and capacitive reactions are both in parallel, so the current flows from one to the other (circulating).

The power of the current that flows through the parallel RLC circuit during resonance is at a minimum. Note: The current flowing is kept in the circuit, and isn't drawing electricity from the source. The glass is filled. The current in the input is at the lowest.

The relation between the phase of current and voltage in a circuit with a series resonance is that the current and voltage are in the same phase. A hint: in an resonant circuit the inductance and capacitance are in opposition so that voltage and current are in sync.

Quantity of the RLC Parallel Resonant Circuit is determined by dividing the resistance by the reactance.

Quantity of the RLC Resonant series circuit can be determined by the reaction time divided in the resistivity.

Memory Point:

Q in Parallel = Resistance/reactance

Q in Series = Reactance/resistance

The half-power spectrum of a resonance circuit that is resonant at 3.7 milliseconds, and Q of 118, which corresponds to 31.4 kHz. Half-power bandwidth refers to frequency multiplied by Q.

Answer: 3.7 MHz/118 = .3135 the MHz number is 31.35 KHz, and 31.4 kHz is closest. There are enough answers from each other that you could be sloppy when you use decimal marks. Find the answer using the right numbers.

The half-power frequency bandwidth of a resonant system that has a resonance frequency of 7.1 Mhz, with the Q value of 150 is 47.3 kHz.

Solve: 7.1MHz/150 = .0473 MHz = 47.3 kHz.

The increase in Q of a series resonance circuit can increase the internal voltages. A tip: increasing Q in a circuit that is a series raises the amount of reactance, thereby increasing the voltages. In a series circuit, Q=reactance/resistance.

The resonance frequency of an RLC circuit in series if R is 22 Ohms, 50 microhenrys are used and the C value is 40 Picofarads., is 3.56 MHz.

The resonance frequency of an RLC circuit that is parallel when it is R is 33 ohms and L is 50 microhenrys, and the C value is 10 Picofarads.. 7.12 MHz.

Resonant frequency is the formula that is

F=1/(2 X LC). If this math is just not enough for you just like I am I have online calculators. I'm not ashamed of providing this tip It's at least in the 80- or 40-meter band of hams.

Increase Q in capacitors and inductors, reduce losses. Note: The resistance that is lost is the reason why the circuit becomes less acute (lower Q). E5B Time and Phase

The word used for the duration needed for a capacitor in an circuit that is RC to be charged up to 63.2 percent of its applied voltage or discharged to 36,8% of its original value can be described as "one time constant."

The time constant for the circuit that has two 220-microfarad capacitances and two 1 megaohm resistors are 220 second. The formula is: capacitance times resistance. Find two 220-microfarad capacitors arranged in parallel equal 444 microfarads. Two resistors of 1 megaohm in parallel equivalent .5 megaohms. 440 x .5 = 220. Make sure to remember the solution is based on the value that the capacitor.

Susceptance refers to the imaginary portion of admittance. It is the measurement of the extent to which a circuit is at risk of conducting an alternating current.

When the size of a pure reaction becomes susceptibility it will become the reverse. Resistance stands in opposition to DC current. Conductance is the opposite. Reactance can be described as how circuits react to AC voltage and current. It is the opposite of reaction (1/X).

The letter B is often used as a symbol of susceptibility.

Impedance can be described as in opposition against AC current. Admittance is the opposite of imperceptibility. The key is that impedance hinders, but admittance lets.

The polarized form of Impedance changed to admittance represents an inverse of magnitude, and changes the direction of the angle. The key to understanding admittance is that it's the reverse of admittance. Find the reciprocal of the magnitude, and then change the signification of the angle.

Impedance is the reason why voltage or current follow one another. In a circuit that is inductive, the voltage is a driver for the current. In a circuit that is capacitive, voltage is lagging the current. Be aware of "ELI the ICEman." The voltage is E, the

Inductance is L, C is current and C is capacitive.

The relation between the current flowing in a capacitor and voltage it produces is that it is the latter that leads the former to 90 degrees. It is true that voltage leads current within the case of a capacitor. Eli the ICEman said so to me.

The connection between the current through and the voltage that flows through an inductor is the voltage leads the current in a 90 degree angle. The key to understanding this is that voltage leads current within an inductor, which is ELI.

The phase angle that is the difference between current and voltage current flowing through the series circuit of an RLC If the XC value is 500 ohms. The R is 1 Kilohm Ohms and the XL is 250 Ohms is 14 degrees, with that voltage being lagging with the current. The hint is that we know that the voltage is not lagging current as the

XC will be more than the XL. This eliminates two possible answers. The remainder of the math is quite staggering. Cheat: The solution to each question can be found at 14 degrees. The circuit is much more than capacitive. Therefore, the ICEman claims that there is an issue with the voltage.

The phase angle that is the difference between tension across the voltage and current flowing through the series circuit of an RLC If the XC is 100 Ohms and that is, the R is 100 ohms, and that the XL is 75 Ohms, is 14 degrees, with the current lagging in the voltage. It is evident that the voltage is behind the current as the XC will be higher over the XL. There are two possibilities. Cheat: The correct answer to each question is 14.

The angle of phase between the tension across the voltage and current flowing through the series circuit of an RLC in the case that XC is 25, ohms, R is 100 ohms while XL equals 50 ohms.

14 degrees, with the current being led by the voltage. It's the same 14 degrees but the circuit is inductive, so the current is led by voltage. ELI.

Chapter 4: Coordinate Systems

Impedances are described using Polar coordinates according to their phase angle as well as magnitude.

The coordinate system that is used for displaying the angle of a circuit that contains inductance, resistance or capacitive reaction is called Polar coordinates. The term "phase angle" refers to Polar coordinates.

Memory point Phase angle = Polar coordinates.

The system of coordinates used for displaying capacitive, inductive or resistive reactions of impedance is rectangular coordinates. Note: If the coordinate system doesn't mention the its phase angle, its rectangular coordinates. The figure E5-1 illustrates this.

The capacitive reaction that is rectangular in notation is the jx. Be sure to examine the voltage of the ICEman. It is a capacitive circuit. The voltage lies behind, -j.

In the case of polar coordinates, a inductive reaction features a positive phase angle. The reason for this is that it's an inductive circuit, so the voltage is higher than the current (positive angle +j).

Diagrams used to illustrate the impedance phase relationships for a particular frequency is referred to as Phasor Diagram. In the answer, if you recognize an example of a Star Trek weapon in the response, then it's correct.

The impedance 50-j25 is 50 ohms of resistance and 25 ohms capacitive reaction. The first is the resistance in ohms. Another number is reactance. Because j is negative we can tell that the voltage is not advancing so the circuit has to be capacitive.

If you use rectangle coordinates to plot the circuit's impedance the horizontal axis will be the component that is resistive. Cheat: Horizontal sounds like reziztive.

These questions are based on Figure E5-1, which is located on the next page:

The number representing a 400 ohm resistor as well as a 38-picofarad capacitor with a frequency of 14 MHz is the point 4. Solution: The answer is on the +400 ohm horizontal line (resistance). We are aware that the circuit is capacitive. As such, it's going to have a negative sign, and will be in the lower part of the vertical the axis. It is at point 4, in default.

A circuit includes a 300-ohm resistor as well as an inductor of 18 millihenry that operates at 3.5 Mhz is point 3. Solve: Inductive Reactance is the ratio of frequency and inductance. 2x3.14x3.5x18= 395. This circuit operates inductively. Find the positive side of the Y axis beginning at 300 ohm along the X axis. The 3rd point is the right solution, as 395 and 300 cross.

The circuit which has 300-ohm resistor and 19 picofarad of capacitor with a frequency of 21.200 MHz is called point 1. Find an axis vertical to the 300-ohm horizontal axis and the circuits capacitive. The solution is in the lower (-Y) quarter. By default, point 1.

Cheat: 300 ohm resistor can be at either point 1 or 3. If the circuit has an inductive component the circuit is capacitive, Y is positive Point 3. In the case of a capacitive circuit it is negative. 1.

E5D AC and RF Energy

The effect of skin is that as frequency increases it is that RF moves through an encapsulated conductor that is located closer to the top of the.

It is crucial to limit lead lengths to components of circuits that are VHF or higher, to minimize undesirable inductive reactions.

Connections that are short can be utilized in microwave frequencies to prevent any phase shifts along the connections. One way is to use the term is "avoid unwanted reactance."

Microstrip is a precision printed circuit conductors that are above the ground plane, which create constant-impedance interconnects the microwave frequency. It's a good idea to simplify it. "Constant impedance" is a great quality.

The direction of the magnetic field that surrounds the conductor with respect to direction electron flow is located in circles surrounding the conductor. Note: If you forget the flow of electrons, instead the magnetic field forms an arc around the conductor.

The real power source in an AC circuit in which the current and voltage aren't in phase is determined using the power apparent by the power factor. If the voltage and current aren't in phase then the capacitor and inductor are feeding one another. Certain power sources recirculate, which means the apparent power will be higher. Add apparent power to the power factor and you can see how much circuits consumed.

The power factor for an R-L circuit with an angle of 60 degrees between the voltage as well as current is .5 Answer: This is the cosine for the phase angle. The cosine for 60 degrees is .5. The test requires an instrument that can calculate scientifically to this test or learn three numbers.

The power factor for an R-L circuit having 30 degrees of phase angle is .866. Solve. The cosine for 30 degrees is .866.

The power factor for an R-L circuit that has a 45-degree angle of phase is .707. The cosine of 45 ° = .707.

Memory Point The higher angle is the lower cosine

Cosine 60 degrees equals .5

45 degrees of Cosine .707

Cosine 30 degrees equals .866

Watts used in an electrical circuit with an efficiency of .71 that has the apparent capacity of 500VA is 355W. Solution the equation: 500 x .71 equals 355 Watts.

When a circuit is designed with the power factor .6 and the input voltage is 200VAC at 5 amps 600 watts will be consumed. Solution: First, transform the volts and amps into watts. P=EI. (P=200 x 5 = 1000. Divide by your power factor .6 600 = 600 watts.

If a circuit is designed with the power factor .2 and the input voltage is 100 VAC at 4 Amperes then the power consumed is 80 watts. Answer: Convert it to Watts: 100 times 4 equals 400 Watts. The power factor multiplied by .2 equals 80 Watts.

Reactive power is not watts as well as non-productive power. It's a inductor and capacitor feeding one another.

The power generated by the reactive energy that is generated in the AC circuit with ideal capacitors and inductors are continuously transferred between electrical and magnetic fields, but does not go away. Tips: If all the components are in good condition and energy does not escape. Check for "not dissipated" in the proper response.

The amount of power used in the circuit is derived from 100-ohm resistors in series with an inductive reaction of 100 ohm that draws 1 ampere equals 100 Watts. Foul! It's a trick problem. This assumes that there is no loss of power in the inductor. What is the power consumed by the resistor part of this circuit? The idea is to not consider the power used by the inductor. It is simply $P = I^2R$. $I^2 = 1$; multiplied by 100. The R (100) is 100 watts.

E6 - CIRCUIT COMPONENTS

E6A Semiconductors

Gallium arsenide can be used as a semiconductor in microwave circuits.

The semiconductor material with excessive free electrons is called N-type. Tips: Free electrons that are too abundant could be charged negativeally.

A PN-junction is not able to transmit current if reverse biased, because the holes of the P-type materials and electrons within the N-type material are separated by the voltage applied expanding the area of depletion. Note: The reverse bias splits and broadens the area of depletion.

The term that is used for an impurity which adds holes to the semiconductor crystal is called an acceptor impurity. Hint: Holes accept electrons.

A DC input impedance of the gates of a field effect transistor is much higher than an bipolar transistor. Field effect transistors feature high impedance at the input.

The beta value of bipolar transistors is the difference in collector current compared to the base current. It is the difference between collector current and the base. Cheat beta and current base.

An silicon NPN transistor is biased at around .6 to .7 Volts. This is because it's affected by voltage and therefore eliminates two of the options. This bias is not large and eliminates the option.

The frequency at which the grounded base the current gain of a transistor is reduced to .7 of the gain that can be achieved at 1 kHz. This is referred to as the cutoff frequency of alpha. The key is that the gain (alpha) is reduced (decreased).

Depletion-mode FETs are FET with an electric current flowing between the drain and source in the absence of a gate voltage used. A hint: Current flow that is not accompanied by gate voltage will decrease.

In Figure E6-1 the symbol that represents dual gate MOSFETs with N channels is #4. The hint is that it's a dual gate, and only two figures have 2 gates (G1 as well as G2). The N channel is the reason the arrow points in. Notice: This is the opposite to bipolar transistors in that they have the arrow pointed out.

The symbol of a P-channel junction FET can be found at the number 1. The P-channel FET is a symbol with the arrow pointed at. The only two symbols to be aware of from figure E6-1.

Numerous MOSFET devices are equipped with internal Zener diodes in the gates, to limit the possibility that static damages will be caused for the gate.

E6B Diodes

The best feature of the Zener diode is its constant voltage drop in the presence of fluctuating the current.

The most significant characteristic of Schottky diodes is that Schottky diode is that it has less than a drop in forward voltage.

Cheat: Find "voltage drop" in both solutions.

An everyday use for Schottky diodes is to serve to be used as a VHF/UHF mixer and detector.

One type of Schottky barrier diode that is a metal semi-conductor junction.

The light source for an LED is a forward bias. This means that it is conducting and going in the direction of forward. The only thing that you should know about Figure E6-2.

A semiconductor device that was designed to usage as a voltage-controlled capacitor is called a varactor diode. It is important to note that the voltage changes the capacitance.

One of the characteristics of the PIN diode which is what makes it a good radio frequency switch is the low capacitance at the junction. In the event of an extremely high capacitance, then the charge could hinder its ability to switch.

For regulating the attenuation and amplitude of RF signals, using PIN diodes, apply the forward DC bias voltage. Make sure to give the signals a boost by using forward bias.

An extremely common usage for a point-contact diode for RF detection. A RF detector can help you make contacts.

If a junction diode malfunctions because of excessive current, the cause is an the high temperature of the junction. A hint: High current can cause excess heat that leads to the breakdown.

The diagrammatic symbol of an LED on Figure E6-2 is 5. It is important to note that the arrows represent shining light from the device. The sole symbol to keep in mind from E6-2.

E6C Digital ICs

A comparator is a gadget that compares two voltages, or currents and generates a signal indicating which one is more powerful. The moment the value of a comparator's input exceeds the threshold, the device alters its output status.

The purpose of hysteresis within an electronic comparator is to block interference from the input signal that can cause instability in output signal. Hysteresis can be described as a delay in voltage between states that are active and inactive. A 12-volt relay turns off at 11 volts. It it doesn't turn off until the it drops down by 9 to. The slightest fluctuations in voltage do not activate it.

Tri-state logic are the logic device that has 0,1 and high impedance output state. It is a hint that Tri-state has three states of output: 1,0 and high.

BiCMOS logic can be described as an integrated circuit family that utilizes both bipolar as well as CMOS transistors (Complementary Metal Oxide Semiconductor). The benefit for BiCMOS logic is the higher input impedance that comes with CMOS, and the smaller output resistance of transistors with bipolar characteristics. The high output impedances are a positive factor because it does not load on the circuit.

The main benefit for CMOS logic devices in comparison to TTL devices is less energy consumption.

Digital integrated circuits CMOS have an excellent resistance to noise in the power or input signal. supply since the input switch threshold is roughly one-half the voltage of power supply. Be aware that it switches for a long time before voltage fluctuations in the power supply can have an effect.

A pull-up resistor or pull-down connects to either the negative or positive supply line to create an electrical voltage in the event that the output or input is an unconnected circuit. The word "pull up" or pull down implies it can be either positively or negatively. The resistor controls the bias.

The Programmable Logic Device (PLD) is a set of programmable circuits and logic gates that form an integrated circuit.

In the figure E6-3, the symbol used for schematics of an NAND gate is number 2.

A NAND gate can produce logic zero as its output when the inputs all have logic 1. Note: N is reversed or negative. The small circle on the output signifies that the output has been reversed. Two inputs form an unidirectional vertical line, they're identical.

Figure E6-3 is a symbol to represent logic circuits. Only need to be able to identify the numbers 2, 4, and 5.

A NOR gate generates a logic zero if all inputs have logic 1. It is important to note that the N indicates that it generates an opposite logic. OR means that either input could be 1. OR signifies that either input can be a number 1.

The symbol used to represent an NOR gate is 4. It is important to note that the inputs have a curved line (OR) as well as the output is a small circle signalling the reversed output.

The symbol used to represent an operation called NOT (inverter) is the number 5. The key is that one input and an output are the minimum requirements for invert.

E6D Torroidal and Solenoidal Inductors;

Inductor saturation occurs when the capability that the center has to accumulate energy is exceeded. It is possible that the core has been overloaded.

The risk of core saturation is to be minimized since distortion and harmonics could be the result. A tip: The core isn't able to keep any magnetic energy and therefore there's distortion.

The equivalent circuit for the quartz crystal comprises motional capacitance and motional inductance and loss resistance together with the shunt capacitor, which represents an electrode, and some the stray capacitance. Cheat: Horrors! It's impossible to be lying. Find the word that has "shunt" in it and get rid of all this.

A component of the piezoelectric effect can be described as mechanical deformation of materials caused through the use of the voltage. The way we think of it is to imagine it in the reverse opposite way: the voltage is generated when pressing a material in similar to the way that a grill light produces sparks.

Commonly used materials to make a base for an inductor include ferrite and brass.

The substance that reduces the inductance when placed into coils is brass. It is important to note that brass isn't magnetic.

The reason to utilize the ferrite cores in place of powdered-iron as an inductor is because the fact that ferrite cores need less turns in order to achieve the same inductance. The key to remember: Ferrite is more efficient

One reason why powdered-iron is used instead of ferrite powdered iron cores retain their features when exposed to higher voltages. The iron core can be able to handle more current since it has the ability to handle heat better.

The property of the material that affects the capacitance of an toroidal inductor is permeability. Permeability is the way in which magnetic flux behaves. It is the measure of efficiency of the central part.

The voltage in the principal windings of transformers with no load is referred to as the magnetizing current. The key is to make sure it's enough to attract the center, but not more as there is no charge to draw current.

The most commonly used devices to serve as VHF and UHF parasitic suppression devices in the output and input connections of a transistor's HF amplifier is ferrite beads. The reason for this is that ferrite beads are also known as toroids. They act as chokes.

The main benefit of having an toroidal instead of solenoidal is that the toroidal one is the fact that it confines the magnetic field in the material of the core. The solenoidal component is a bar of ferrite that has the wire covered with. The toroid's circle helps to keep the magnetic field in check.

The main reason for self-resonant inductor behavior is the inter-turn capacitance. The wire turns interact in the form of capacitors which create an resonant circuit along together with the inductor.

E6E Analog ICs; MMICs; IC Packaging

Gallium arsenide (GaAs) can be useful in semiconductor devices that operate at higher frequencies and in UHF due to its higher electron mobility. It is important to note: Electron mobility is a odd enough word to remain with you.

One example of a through-hole device is the DIP. Dual Inline Package is one of the types made of chip for computers. Leads are soldered into holes that are in the board circuit.

MMIC is a reference to Monolithic Microwave Integrated Circuit.

The substance that is most likely to give the greatest frequency of operation for MMICS

MMICS is gallium Nitride. The hint: Gallium is employed at extremely higher frequency. Gallium can also be the solution to a different question.

The standard amplified output and input impedance of circuits using MMICs is 50ohms. It's similar to coax.

The standard noise figure of a preamplifier that is low-noise is 2 decibels. A hint: the pre-amplifier creates certain noises, however, not much. Other answers can be negative or are too high.

It is able to supply power to an MMIC via a resistor or RF choke that is connected to the lead that is connected to the amplifier's output. It is important to note that power comes from the lead that is output.

One of the features of MMIC that makes it a popular choice in VHF via microwave circuits is the controlled gain as well as a low noise figure as well as constant output and input impedance across the frequency band. The best part is that you don't have to be aware of everything. Pick "low noise."

The line of transmission used to make connection to MMICs is called microstrip. The key word is microstrip.

The most appropriate component packages for applications in the higher frequencies is surfaces mount. The surface mount package has the longest leads (none) as well as high shorter leads minimize stray reaction in high frequency.

The benefit of using surface-mount technology over components with through holes is

* Circuits with smaller areas

* Lesser circuit board traces

* Components with less parasitic capacitance and inductance

These choices are true.

The trick is to recognize two and then you'll know that the solution is simple.

The most distinctive feature that is characteristic of DIP packaging is the two rows of pins connecting in opposite directions of the packaging. Hint: Dual Inline Package.

Through-hole package ICs with DIP aren't typically utilized at UHF and other frequencies because due to the excessive length of lead.

Chapter 5: Optical Components

Electrons absorb the energy of sunlight that hits an photovoltaic.

If light hits any photoconductive substance, its conductivity rises. Hint: Photoconductive.

The most popular configuration for optocouplers or optoisolators includes an LED as well as a phototransistor. Two circuits can be isolated (not electrically linked) However, the signals are transmitted through the light.

Photovoltaic effects are the conversion of light energy into electric energy.

A optical shaft encoder senses the movement of a controller through the interruption of a light source by using a pattern wheel. In the event of a shaft turning, it detects flashes of light as the shaft rotates.

One of the most common materials employed to build photoconductive devices is called a crystal semiconductor.

Solid-state relays are device that makes use of semiconductors to accomplish the tasks in an electrical relay. The key to understanding a solid-state relay is that it is a relay constructed from the use of solid-state electronics (semiconductors).

Optoisolators are commonly used when working with solid state circuits for switching 120VAC, because optoisolators offer an excellent level of electrical isolation from the controller circuit to the circuit that is being switched. Hint: Optoisolators provide isolation. Search for the word "isolation" in the solution.

The power of a photovoltaic device is the proportion of light which is converted into electricity. It is important to note that the more efficient is the cell, the greater amount of energy it can produce with a specific quantity of light.

The most commonly used kind of photovoltaic cells utilized for power generation is silicon. It is important to note that solar panels are constructed out of silicon.

The average voltage generated by a fully lit silicon photovoltaic device will be 0.5 V. Every cell generates 0.5 voltage. Connect the cells together to produce a greater amount of voltage.

E7 - PRACTICAL CIRCUITS E7A Digital Circuits

A bi-stable circuit can be described as a flip-flop. The key word is "Bi" means it has two states: flip and flip.

The primary function of a counter IC is to create one output signal per 10 input pulses. In other words, it counts either decades or 10s.

A circuit that is able to divide the pulse's frequency train by two is called a flip-flop.

For dividing a signal's frequency into 4 parts, you need 2 flip-flops. Hint: 2 X 2.

An electronic circuit that constantly changes between two states, without any external clock is known as an astable multiplevibrator. Note: Astable signifies that it is constantly changing. The system is not stable.

Monostable multivibrators switch momentarily into the other binary state before returning to its state of origin after the set amount of period of time. The word monostable is used to describe how it returns to a single state.

A NAND gate creates an underlying logic of zero when all inputs are the same logic. It is important to note that N creates an opposing logic. And that means both inputs have the same logic.

An OR gate generates one if all or all inputs have logic 1. It is important to note that there is not an N before the gate, therefore output will equal input. OR indicates that either input could be 1.

A NOR gate creates logic of 0 when only one input is logic 1. It is important to note that the N indicates it creates an opposite logic. OR is the opposite of N. OR implies that one input could be 1.

A truth table can be described as an inventory of inputs and equivalent outputs of an electronic device. A hint: It's one of the tables showing different possible outcomes.

The logic behind defining "1" as a high tension is known as positive logic. It is important to note that 1 is a number that can be considered positive.

E7B Amplifiers

Amplifiers are able to operate across either the entire or a portion of all-round signal cycles. Their class defines what percentage of the cycle runs.

The push-pull Class AB amplifier conducts more than 180 ° but fewer than 360 degrees.

An amplifier of Class D makes use of the technology of switching to attain an extremely high level of effectiveness.

The elements of class D amplifiers comprise a low-pass filter that eliminates switching signal components. The switching circuits may create spurious signals. The low-pass filter eliminates these signals.

The load line in an Class A common emitter amplifier will generally be biased between cutoff and saturation. The Class A line runs across the whole 360 degree cycle therefore it needs been biased to the middle.

In order to prevent unintentional oscillations from an RF power amplifier put in parasite suppressors, or eliminate the stage. Tips: suppressors block unwanted oscillations. Neutralization involves introducing negative feedback that cancels the oscillations.

The RF power amp can be neutralized through feeding the 180-degree phase out of its output to the input.

An amplifier that decreases or eliminates harmonics of even order is known as a push-pull model.

If you use a class C amplifier is employed to amplify SSB then you will experience significant bandwidth and signal distortion. Class C functions only about 50% of the duration. Class C isn't linear, and it is best suited to CW however is not SSB.

Tuning a vacuum tube RF power amplifier which makes use of a Pi network output circuit, you must adjust the tuning capacitor for a minimum plate current, and also the load capacitor to ensure that it is able to handle the maximum current. Tips: Set the tuning capacitor at a minimum level and load the circuit to maximize.

Figure E7-1 shows a typical transistor emitter amplifier.

Three questions are related to figure E7-1. To answer them, you just need to recognize the figure by its designation as "common emitter" and know the reason behind R1/R2 and R3. The purpose of Figure E7-1 of R1 and the R2 are to set a fixed bias. They control the voltage at the bottom of the transistor. They are a type of voltage divider between ground and +.

The reason for R3 is to provide self bias. The purpose of R3 is to bias the whole transistor. Amplifiers are operated by signal inputs which range from negative to positive. This is a way of establishing the proper operation point for the transistor, and, when performed correctly, decreases distortion.

A emitter-follower (or commonly referred to as a common collector) amplifier is a device with a low impedance output which will follow the voltage at which it is input. Hint: An emitter follower follows.

They are much more effective than linear amplifiers due to the fact that the power transistors are in saturation or shut off most times. One way to look at it is that is to say it conducts only for only a small portion during the entire cycle.

To avoid thermal runaway within a bipolar transistor, you must use an in-line resistor to the emitter. It limits the amount of current that may be flowing and can cause the transistor to overheat.

The effects of intermodulation of the linear power amplifier is the transmission of signals that are spurious. It is important to note that intermodulation can cause distortion, and it can cause false signals.

Intermodulation of odd-order, not even-order distortion products pose more problem for linear power amplifiers since they're relatively near to the signal desired. The fact that they are close to frequency can be a challenge.

One of the characteristics of a grounded grid amplifier is its low input impedance. In the event that you see that the grid has been grounded it's at a an impedance that is low.

E7C Filters and Matching Networks

Inductors and capacitors in the low-pass filtering Pi-network are set up by connecting a capacitor between ground and the input and a second capacitor connecting both the ground and output, and an inductor between the output and input. A hint: There is a more complicated answer. The purpose of the filter is to transmit low frequencies.

Therefore, it needs to divert high frequency signals. How to accomplish this is to use capacitors that ground the output and input. You can look for them as a one of the answers.

The T-network's characteristic comprising series capacitors and the parallel shunt is an inductor that is a high-pass filter. The reason is that series capacitors can are able to pass high frequency, so they are a high-pass filter. This is the opposite of what we have said above.

A network with a Pi-L has advantages over Pi-networks due to its higher harmonic suppression. A hint: the extra component (L) gives more protection.

A circuit that matches impedance transforms the complex impedance into the resistive part of it by cancelling that reactive component of the impedance, and then changing the resistive component to the desired amount. The impedance matching process transforms into a desired value. Search for "a desired value" in the solution.

The filter that has an amplitude in the passband, and the sharp cutoff, is known as Chebyshev. Cheat Recall Chebyshev as a Russian name. This is the only solution.

The distinctive features of an elliptical filter is very sharp cutoffs, that has one or more notches on the band that stops. The trick is to say: "Extremely sharp cutoff" is a good filter, and so is the solution.

The Pi-L network that is used to match the vacuum tube's final amplifier with a 50-ohm balanced output is an Pi network that has an extra series inductor at the output. The Pi-L network adds the letter"L (inductor) to an existing Pi network. It's all you need to be aware of.

The most significant factor that has an influence on the bandwidth and the shape of the crystal ladder filter lies in the frequency of each crystal. It is important to note that the crystal frequencies define the design that the filter will take.

A crystal lattice filter an instrument with a sharp skirts, a steep skirt and narrow bandwidth constructed from quartz crystals. Many crystals are linked together in a lattice which makes the filter sharper and more narrow. It is important to note that the crystals are not in a lattice shape; instead, it's a circuit.

The ideal choice for a filter for a 2 meter repeater is the cavity filter. Cavity filters comprise the primary components of a duplexer's circuit and are resonant, sharply tuned circuits which allow certain frequencies to go through. Duplexers separate its receiver and transmitter when share the identical antenna.

Chapter 6: What Is Ham Radio?

In this article, we'll explain exactly what ham radio is, sometimes referred to amateur radio. We will also discuss as its history, the way it got its start along with a bit of information about everything else that pertains to this topic.

In the remote town of the ex- Soviet Union, a youngster remembers the first time they made contact with a nation in South America using their own Ham radio, which was designed and assembled by the designer. It was the middle of the 1980s and strict rules set by the Soviet system was to reduce the data reported such as the name of the city, its operator, signal received and temperature local to the area. This is a significant technological breakthrough that will allow us to communicate with the universe that would never appeared possible without.

A long distance from home, a seasoned South American amateur radio operator starts contact through the communication records books (QSO) and then begins to sketch in your brain what the life, culture and daily life of his Soviet partner really are.

This scenario appears to be inconceivable in our current age, when communication can be instantly via the Internet, mobile phones satellites, as well as global TV networks. Yet, in our contemporary times, each day millions of amateur radio users are still able to connect with each other around the world, forming a huge world-wide network of friends. it's social networking at its finest and an excellent opportunity to connect and get to know people that you may not otherwise have met.

More simply Ham radio, also known as amateur radio is one of the most popular hobbies However, it is also able to be employed in more formal settings, e.g. aiding in emergency situations even when other communication methods are not working or damaged. The 9/11 attacks were a prime example of ham radio. Ham radio was instrumental in coordinating emergency rescue operations, which illustrates that hobbies are also extremely beneficial and, in some cases they can assist in saving lives. At a basic level, however Ham radio can be used for bringing people together individuals from every walk of life. This can be enjoyable social, enjoyable, and educative.

A definition of an amateur radio operator is someone who for fun utilizes a ham radio to communicate, without commercial purpose or with people who are also engaged in the same activities. Based on the type of equipment you use the communication could occur on your block or international, or even with an amateur astronaut from the International Space Station - basically it is possible to communicate with anybody in this manner. Ham radio communication can be conducted through voice using digital methods, or making use of a computer.

Some ham radio operators prefer using the most traditional method of wireless communication, Morse code, or telegraphy which is a personal preference which they've become comfortable and familiar with.

Ham radio is the operation by amateur radio transmitters that communicate with other people for a range of motives. The word ham in ham radio originates from the negative term that was used to denigrate amateur radio users throughout the late 19th century. The insults 'ham-fisted', or "ham actor" were directed towards and later used from the amateur radio group as well as used to refer to themselves to be the movement's own.

An authentic amateur radio operator is also who is interested in scientific and technical things, who is able the idea of conducting experiments with antennas, equipment and assemblies. The most modern communication technology, including mobile phones as well as other advancements in technology are available to everyone because of ham radio, which enabled these technology to be developed and tested thoroughly.

Many people who become involved in the ham radio industry are attracted to the radio technology to enhance their own lives to pursue their personal interests. This is that radio communication across very long distances is an old-fashioned technique of interaction with other people but the internet is more effective and is advancing to greater standard.

Then why do there seem to be the vast majority of people running Ham radio stations all over the globe?

The ubiquity of ham radio may have some connection to its significance in the past of interaction between humans. The introduction of radio communications during the 18th century and the early 19th century paved the path for succession of technological advances that were massive that led to the technologies we use available to us today. The benefits would be immense anyone who was able to be a master of radio communications in battle as this was acknowledged during the time. Many wars have been won because of the capability of one side equipped to utilize radio communication superior to the other as well as methods of communication during emergencies were improved because of radio. You can clearly see that radio goes beyond than just listening to music or the information.

Ham radio played, through its existence, played an extremely crucial role in aiding of disasters, situations of solidarity as well as public emergencies around the globe. In the past, when cities weren't connected to phone networks It was typical to have local amateur radio services assistance to distant relatives to obtain medications that could only be obtained in major cities or at a foreign location. Many lives were saved due to the comradery of radio amateurs to obtain medicines that were not available in the local area. For instance, to illustrate one aspect that shows the importance of this particular activity for communicating in times of crisis and emergency situations, one American amateur radio operator took part in an emergency communications network in the aftermath of the terrorist attack on the 11th of September. This helped to help save lives. This can tell you lots about the ways in which ham radio communications can be utilized in a variety of diverse ways.

Consider all the ways you utilize a radio throughout your day; it is possible that you listen to FM radio in your car. And there's a chance that you have a preferred radio station that is located in an exact place on your dial. it's a typical instance. If you've utilized a cordless telephone at home or garage door remote opener, you've probably used the most basic radio communications technology before, perhaps not even noticing that you had it.

For instance, if you want to use an opener for your garage, you press a button which connects to a battery located inside of your remote handheld Then, power is transferred to a transmitter which broadcasts the exact signal to all devices may be within the the range. The motor connected to your garage door will begin to function when it gets the signal transmitted via the remote.

Since radio waves move directly through almost all substances, with the exception of metal, it's relatively simple to keep a clear reading of a radio signal, even if you're in a structure or through a tunnel along the highway. Communication methods through amateur radio use the same format for reception and transmission of radio frequencies These two words simply translate to "sending" and "receiving.'

Some gadgets can send and receive signals as well, which is why are referred to as "transceivers'. If you are tuning into an FM station with your radio, you aren't transmitting any signals; instead, you're instructing your radio to accept signals that are broadcast on a particular frequency. In this case, 99.7 FM refers to the frequency that every song and talk for the station are broadcast. it is also the frequency that is the frequency you choose to tune into your radio in order to gain access to the content.

However, it's more than just listening to what another is transmitting but eventually, the majority of Ham radio operators wish to broadcast a signal that others can listen to - they would like to broadcast themselves in order to connect with each other in a larger method. In most cases, they need to get special licences from the government that they operate under in order to broadcast their message within the bounds of law. And it is a requirement to be sure to do when planning to broadcast yourself.

Radio transmissions are typically monitored by the national government since they're utilized to broadcast emergency messages as well as military and many other important purposes. Thus, many governments wish to ensure that everyone that is operating a radio transmitter that is at a higher wattage is aware of the procedure they're using and the frequencies they are making use of.

This is one example of what can be the outcome if radio operators had a bad attitude - Maybe there are a number of traffic junctions operating with timers which communicate wirelessly with one another. it is an incredibly popular method for automating stop lights as well as various traffic signal systems. If a radio company was to begin broadcasting signals using the exact frequency which is utilized by timers that are connected to the lights at the intersections of traffic the result could be an unintentionally dangerous error because of a miscommunication maybe without both parties being aware of the issue prior to.

The United States, it is the Federal Communications Commission, or FCC which is accountable to oversee all radio transmissions that occur throughout the nation. Radio frequencies accessible to consumers differ widely, and some have been utilised for particular purposes. It is the reason why the license will grant users the right to utilize any frequency that is available, with no conflict or problems.

ISM refers to a range of frequencies which are "license-free" - ISM is a reference to medical, scientific and industrial. Within these ISM bands are 900 MHz radio frequency as well as a few 400 MHz radio frequency. But it's not only the frequencies which are strictly controlled by government agencies, because there are limits to the maximum power - also called output power which can be utilized for any particular broadcast. This is due to the fact that powering up another radio system is extremely likely if someone chooses to put a large amplifier to their broadcast and could disable another radio channel. It's really about everyone having a fair and accountable relationship with the other.

As each country has unique regulations regarding radio It is recommended to conduct your own research before you start. Consult with the local authorities on the radio frequencies permitted to use as well as at what power ratings. It's also best to be involved with local clubs for ham radio that are comprised of members who have a good understanding of the subject They can provide great source of information as you are learning about this fascinating hobby as well as meeting new similar-minded individuals.

The hobby that demonstrates diversity

If you're not a radio amateur You don't know what a variety of unique and fascinating things you can accomplish through this pastime. Who can be in contact with? Which new contacts will you meet? As you wander in the streets of your town, you are sure to meet individuals of every kind, both women and men, people from different age groups as well as different ethnicities, social classes and religious beliefs. They might be housewives, engineers police officers, drivers or bank clerks and anyone else. Any among them may become a amateur radio operator that you accidentally keep contact by radio.

Radio amateurs are an activity that is democratic and is not tolerant of discrimination on the basis of social in any form, whether racial or otherwise. The issue is not important to an amateur radio user if their friend on the other end is not of similar beliefs or political views, or in the case of another race. Ham radio is a world community that has divergences in a variety of ways however what is important is that everyone shares the same interest.

If you are an amateur radio operator there are a lot of possibilities of hobbies and interests. Amateur radio operators are those with the Technician Class license. They have a sole purpose of speaking locally on the VHF and UHF bands within 65 and 130 miles. Some prefer operating with HF devices that allow contacts that extend to hundreds of miles. A lot of amateurs enjoy riding their own antennas, playing on new antennas and circuits. Some even install their own antennas.

Low power radio is known as QRP call and it is a fascinating concept for those who see power limits as an obstacle. Talking to others by using Morse code is still a fascination for millions of amateur radio enthusiasts all over the globe and users who want to create contacts using RTTY (radioteletype) and digital mode using a single radio and to a computer. Ham radios also allow contacts through satellite (yes there are exclusive satellites specifically designed for radio amateurs) and contacts via moon's reflection (the signal is folded onto the surface of the moon) as well as stay in contact via ham radio on Space with International Space Station (most astronauts are amateur radio operators). This is the reason why the ham radio its popularity, its reach is so broad.

If you are interested in other activities in addition to ham radio, like the idea of off-road rallies, navigation or you enjoy biking trails, then you should consider giving your car the latest ham radio technology to provide greater safety, and companionship on long journeys.

Ham radios have an ethics code that was put together in 1928 by American Amateur Radio Operator Paul M. Segal W9EEA:

1. Consideration - It is important that the Amateur Radio Operator is aware and does not allow his station in a way that harms other activities.

2. Loyal It is the Amateur Radio Operator is loyal and provides their loyalty, motivation and assistance for their fellow radio enthusiasts, their club at home and their local club and the American Radio Relay League, through which Amateur Radio in the United States is represented both nationally as well as internationally.

3. Progressive Progressive - Amateur Radio Operator is progressive and ensures that its station is modernized, updated and maintained correctly installed and functioning smoothly.

4. Friendship - The Amateur Radio Operator is a good friend and is patient with his colleagues, particularly if they're just beginning to learn. They offer advice and assistance to novices, while also offering

support and cooperation. They also consider collaboration with those of foreign interested parties, and this is characteristic that make amateur radio a unique form of radio.

5. Balanced - The Amateur Radio Operator is balanced. Radio is his pastime and he is not going to permit his passion to hinder all of their duties, chores at home, work and school, nor the society in which he lives.

6. PATRIOTIC: His place and abilities are always at hand to serve his nation and the people of his area.

Contests

Competitions are held that are open to amateur radio operators similar to what Olympics are held for athletes. It is an opportunity for displaying talent and skill and also an opportunity to continuously increase the quality of conditions, technological and operational.

A rise in operating technique as well as improved efficiency in operation is the main outcome for an operator that will take part in competitions whether an avid competitor or even a participant. If you're not sure regarding this, you can listen to the music of any band or all of them, and listen to the most effective operators. It is likely that the most of the people that have the best quality and are constantly participating in competitions.

The contests manager knows by experience that being condensed and short is vital. Also, the contests host likes having one of the most effective signaling options available in the band. The station may not have the best equipment nevertheless, an indication of the efficiency of using components at the station's equipment is crucial.

Participation in contests is an effective method to connect with local stations who must work hard to be eligible for a variety of degrees. One aspect to be aware of is that contests with serious operators are not a fan of delay, therefore don't declare during the contest that you require to provide the QSL and then cite it on the QSL card you provide. In fact, it isn't necessary for you to have an Azerbaijani (in either CW or telephony) for contests or perform DX. You just need to be sure to stay clear of the upper boundaries of QRG's in which most activity is usually concentrated.

After you've mastered the basics of radio technology, its uses, utilized and what the motives are it is used, let's dive further into a few fundamental concepts that are used of the field of ham radio. Be assured that you won't be overwhelmed with the amount of knowledge, or haven't mastered some aspect of it, since we'll explain in detail what each one means and provide an easy-to comprehend guide to make sure you are aware.

Chapter 7: Ham Radio Made Simple

So, in the previous chapter, we went through the basics of ham radio and introduced you to the fundamentals, in addition to the various applications it is used to serve. The chapter introduced quite a bit of knowledge in our previous chapter, which isn't easy to comprehend. This is a complex matter if you're not aware at the beginning, and this is why we continually reiterate the information we've discussed, in order to help your mind more at ease with basic concepts of the field in addition to giving the necessary confidence to go into the future and take on the game in your own way.

It is clear the meaning of ham radio however, what exactly is a typical ham radio user appear like? What kind of day-to-day task do they have? What is the main draw?

Let's talk about it in plain English.

What do Ham radio operator appear like?

Then, the person you're looking at is exactly the same as you!

There's no single size that is perfect for all looks or ways of life for a radio operator. Just like there's no single size that to fit all manner of living for those who play football or someone who loves reading cooking as an activity, or who enjoys crafting activities A hobby is pastime.

Ham radio operators are from a variety of types of backgrounds, many different nations, female and male, wealthy and impoverished, professional or those in low-level professions. just an interest in the field of communications and the desire to interact with other people.

In addition Ham radio operators are able to explore the depths of their radio, with modern equipment or operate on a simpler scale. It depends on how much money they have available and how they'd like to get.

What's the huge appeal of radio hams?

Ham radio has made an appearance on Hollywood and this tells you the extent of its appeal! Watch The Last Man Standing, the ABC TV show, and you'll find Tim Allen becoming au fait of amateur radio.

What is the attraction?

An opportunity to connect with others across the globe, with the same interests as you do which is the main reason. In our previous section about the ham radio helping in times of disaster as well. That's another important reason to consider it, as it's always at the back of the line ready to leap into the fray whenever needed. However, on an overall note Ham radio is a great way to meet new people, and proving that the world really is larger than you would have thought.

What are the ways to connect with other ham radio users? Simple!

The vast majority of information that we will discuss when we get further into this book may seem complicated at first look through it. That is because you might not have a lot of knowledge about this subject right now but rest be assured that this is going to change in the future, and as you get more familiar with the subject it will become clear that the concepts and definitions will be easy to comprehend, and simply, straightforward. But, for a basic point of contact, how ham radio operators interact with one another is largely dependent on the technology used.

Voice communication can take place in conjunction with microphones, or it may be via messages. It is possible to use Morse code, if it is something that you are interested in however, you do not have to understand the concept completely. It is possible to communicate with people from across the globe, or could even talk to your friends at International Space Centre via satellites within space. If you're looking to get technically proficient, you could bounce signals off of moon!

The key is to experiment at beginning, and then understanding the fundamentals of radio communications - this is a topic we'll cover with greater detail in a subsequent chapter since it's an important factor to remember, especially in dealing with experienced radio users.

Many lifelong friendships made through amateur radio. In fact, it is believed there are 2 million amateur radio people who use radio around the world It's impressive, isn't it? !

Do you require a licence or permit?

This is something we'll discuss in greater in depth when we move through this guide, as it is true that you need to have a permit, and the absence of one could put you in hot waters. Making sure you have the proper paperwork in place from the beginning will spare you from a ton of headache in the future.

In essence, at present the moment, we'll say it is true that you need to pass an examination that is recognized by the FCC which permits users to utilize the ham radio for communicating with others operating on amateur frequency bands. In addition there is a need for an understanding of radio technology and the principles behind it as well as the fundamental rules of using the sport.

Does ham radio cost a lot of money as a hobby? It is as expensive as you'd like to make it in many ways. There is an established cost for the test and materials for study before you are able to begin getting yourself in the air, however it's a small amount - approximately $40 which includes your licence. Once you've achieved it, the cost will start!

Equipment will provide the most output when it comes to money However, you can get it cheaper through shopping in addition to buying used equipment. You will come across numerous radio flea markets. These are where you can get bargains. If you're unsure how to begin in the context of this, ask assistance from fellow amateur radio operator and also check out online forums.

The gear you purchase will most likely increase as time passes, and you'll be able to add more it when you get acquainted with the field. In general it is possible to purchase your equipment for 200 dollars or less if purchase second-hand, which is an excellent starting point, before you figure out if Ham radio is right the right choice for you.

What is the reason it's called"ham radio? What is the reason operators are called Hams?

It's gray area, but it is generally accepted that the whole thing dates back to the first wireless radio users. The air was active during this time as many operators tried to talk with one another in order to perform work from one side of the city to the other there were delays due to the operators that would speak too often about things that were not or too significant - they were called Hams since they essentially were the ones who spoke what they considered to be nonsense.

The phrase has remained in use until today in that it's frequently used, without any thought about the meaning behind it! Ham radio operators using their names openly as hams, despite the fact that the original explanation for the word was not very favourable!

How can I tell if Ham radio is the right choice suitable for me?

In simple terms the answer is that you will never know until you attempt!

A lot of people begin their journey on ham radios with only an interest that is not too strong, and they are conversing with other operators, studying about it through the Internet or in books and generally speaking in websites about it to people who might be interested. This is a fantastic method to experience the hobby and get an idea of the hobby without the expense of buying equipment you may not utilize. However the ham radio hobby is extremely addictive as soon as you get started to connect with others across the globe You will soon realize it hard to resist making contact with many increasing numbers of people.

It's important to note that it is important to know about proper conduct and the rules and regulations of Ham radio prior to you engage in communication with others However, the majority of the time it's just normal sense and the proper manners.

Naturally, there's the code of conduct like with everything that involves people, however it is easy to learn and observed.

The next step is to look at some more specific aspects of the Ham radio that are crucial to be discussed to enable you to fully comprehend how it all functions. If you are lost, or feeling as if you were stepping into the dark side you can simply go back and read this book again. feel secure knowing that we'll be reiterating the information over and over again throughout the book to make certain that you comprehend what you're doing.

What is the ARRL?

You may encounter this idea when you are looking into the world of radio ham, and certainly it's worthwhile to learn more about it.

It is the American Radio Relay League (ARRL) is a radio association in the US in the US, as one would think, and is a nationwide association that promotes Ham radio. As of today, there are more than 16,000 people who are members of the organization This means that it is an opportunity to find out more about it, and to network with new fellow radio operators. They are exactly like you!

Information provided by the ARRL is up-to date as well as relevant. You are able to access information, websites courses, books, and classes for more information. The ARRL has educational classes offered across the United States for those who reside in the US this is) great opportunity to gain knowledge and receive advice as well as) an excellent opportunity to inquire about subjects which you're not sure about as well as is) an excellent social occasion for those who are similar to you to connect.

The chapter ought to have addressed the majority of concerns you may want to be aware of about ham radio and also those annoying concerns you may have thought of asking someone other than you, for fear being embarrassed! Do not be worried about this when it comes to Ham radio because nearly every question you can think of has probably been asked previously and experienced operators are more than eager to help with whatever issues or concerns you might have. The development of ham radio is a major goal for any operator. Therefore, assistance is definitely available to help you to further pursue your passion.

Chapter 8: Key Concepts

The first introduction to basic notions by delving into a few of important terms thought to be fundamental to the process that is radio-based data transmission. The focus will not be solely at the effects of artificial radio signals as well Did you know there is a possibility that radio frequency (RF) emits are natural? We'll discuss that in the future to provide you with more knowledge of all radio spectrum.

What Is a Radio Wave?

Perhaps you noticed that I've used the word "RF" in the preceding paragraph. This is seen several times throughout the book to refer to any aspect that is related to radio-related information, regardless of whether it's an identify beacon, a fragment of spoken human language, or just the sound of a radio.

"RF" is used to describe any kind of electromagnetic radiation which falls within the frequency range between 3 kHz and 300 GHz. When discussing frequency bands, it's essential to understand that we're talking about something that is known as the electromagnetic spectrum. This is an expression which refers to the entire spectrum of frequencies for electromagnetic radiation. It is interesting to know that when we go to a certain point within the EM spectrum, we will arrive at visible light. This is in reality a thin EM section, which is located between near infrared and ultraviolet. Confused? Don't be! All will be transparent.

Technically speaking what we see as visible light is visible to our eyes is comprised of the same type of radiation that radio waves used for radio communications. Since we're focused on EM radiation that is between 3 kHz and 300GHz We won't discuss frequencies beyond this range. When we talk about the kHz frequency and GHz...let's start by delving into what these terms mean also.

Hertz

Heinrich Rudolf Hertz was the first to demonstrate that electromagnetic waves exist which is why his name was utilized as a unit of measurement used for EM radiation. Hertz is used now for 'one cycle every second' therefore a light blinking for a specific period every second is thought to operate at the rate of 1 Hertz. If a light flashed just once every 2 seconds, would you be able to guess would the Hertz value could be? It's .5 (half of one Hertz) It's isn't as complicated than you thought!

The Hertz measurement isn't just limited to EM spectrum phenomena. It is also applicable to areas of vibration measurement as well as computers. However to the purpose in this paper by using the word Hertz means, we're talking about the unit of measurement that has been relegated in that of the EM spectrum. Below are some prefixes that are commonly used as well as their Hertz rating:

Val SI symbol name SI symbol Name

10^{-1} Hz dHz decihertz 10^1 Hz daHz decahertz

10^{-2} Hz cHz centihertz 10^2 Hz hHz hectohertz

10^{-3} Hz mHz millihertz 10^3 Hz kHz kilohertz

10^{-6} Hz uHz microhertz 10^6 Hz MHz megahertz

10^{-9} Hz nHz nanohertz 10^9 Hz GHz gigahertz

10^{-12} Hz pHz picohertz 10^{12} Hz THz terahertz

10^{-15} Hz fHz femtohertz 10^{15} Hz PHz petahertz

10^{-18} Hz aHz attohertz 10^{18} Hz EHz exahertz

10^{-21} Hz zHz zeptohertz 10^{21} Hz ZHz zettahertz

10^{-24} Hz yHz yoctohertz 10^{24} Hz YHz yottahertz

Electromagnetic waves propagate themselves as the repetition of photon pulses. The pulses are detectable and have frequencies that permit us to determine their frequency based on how fast they move. That's why, when you set your radio to a particular radio station, you're making sure that your radio's RF receiver is tuned to the precise oscillation speed of the station which plays the tune that you want to hear If you're off-kilter, your radio won't play through the speakers or will sound unclear or off.

At a basic level that's the way radio communications work. Before we can begin making use of radio technology, we require a basic understanding of other key terms, in order that we're not totally disorientated when it comes to construct our very initial radio set-up for amateur radio.

RF in Other Daily Uses

It is possible that you have heard "RFID" as an acronym prior to that. It's a shorthand for Radio Frequency Identification which is now widely used to monitor data remotely stored in or stored on an RFID chip. It works by making sure it is active or passive RFID chip which is fitted on something that needs to be monitored.

One good example can be seen in RFID toll cards that are utilized to pay for tolls on roads without needing to stop or slow down in order to cash out physical cash. The reason for this is that there exists an active RFID chip connected to a battery that can be found in a small container that is placed on the windshield of the driver. If this tiny box comes in contact with a radio receiver of what's called "interrogatory radio waves," it becomes detectable immediately and can be examined for its underlying data.

If you have an RFID chip surgically implanted in dogs, the chip is not active and doesn't include batteries. But, it can be read within a short distance using a powerful radio transmitter. That's how animals that have been lost can be reunited with their owner Do you realize the significance of radio communication today?

Do you realize that the home microwave, which that you use to cook your food items, can be the source from EM radiation? How about the sun? It also emits huge quantities of EM radiation. But only a tiny fraction (a extremely, tiny portion in reality) can be observed by the naked eye. Of course, we are not able to gaze directly towards the sun!

Globally, Earth Earth is continually bombarded by EM radiation, which strikes our planet from all angles and originates from stars within our galaxy as well as all over the universe. Why doesn't this vast ocean of noise not squelch the signals we receive? The answer lies in an aspect known as amplified.

Amplitude refers to a word that can be used to define the strength of radio waves as well as how strong it can be. Radio signals that are broadcast with just 1 milliwatt is very difficult to detect compared with the same radio signal which is broadcasted at four watts, or 400 Watts.

The word "watt" may be well-known to you as the measurement of power which is not an accident that in order to send any kind of EM radiation at all you must find an energy source. For the sun, it's nuclear power. In the case of walkie-talkies the power source is battery coming from, for instance, an 9-volt battery in each phone. Each and every thing that sends an electronic signal falls within the same category of understanding and all of it results in some form of communication.

For amplitude to be added in a radio signal the power source can be applied to any antenna that is being utilized for transmitting the signal. The topic of antennas, amplified signals and increase in the coming chapter, so don't be concerned that you're struggling to grasp this concept right now. In the meantime take note of these terms and be aware the importance of them in the overall scope of radio.

This chapter was designed to provide you with an overview outline of radio frequency as well as electromagnetic radiation. In chapter 3 in which we'll begin to look at the subject matter at a higher level. Although there are numerous words to become familiar with, they'll come to be second nature of knowledge as we go through the book. In addition, we'll go over their significance over and over.

Radio Frequency Spectrum

The word "frequency" is one you likely have heard time and often in relation to radio. Usually, it's something basic such as changing the frequency of your radio the station of your choice. Frequency can be described as a phone lines in a variety of ways. It is an essential communication channel as well as the channel's spacing is the variation in frequencies between two stations. It is typically 12.5 KHz as well as 25 KHz. But the frequency range of radios is directly related to the power of it as well as the geographical location of its location as well as whether or not it has repeaters, not its frequency. When it comes to RF output power can be described as what is known as the "power" of radio, which determines (amongst other factors external to the radio) the range of radio. It's determined in watts. In order to make it easier for calculations each watt equals around one KM of range.

The total frequency spectrum of radio is split into four parts. This division is required due to the fact that waves with identical frequencies may have identical properties. This are described as

1.) Higher Frequency (HF) is also referred to as shortwave. These are frequencies that fall below 30 MHz.

2.) The Very High Frequency (VHF) encompasses the spectrum between 30 to 300 MHz.

3.) Ultra High Frequency (UHF) The range of 300 MHz up to GHz.

4.) Microwave: This is the frequency above GHz.

A Basic Rig

The knowledge of all these subjects is helpful, but in itself it won't grant an entry into the world of Ham radio. In order to do that you'll need at a minimum an initial radio set-up and we'll get through in greater detail soon.

The basic configuration is commonly called "rig" by ham operators. It consists of a receiver as well as a transmitter that allows users to hear others operating and converse with them.

The other important aspect of the equipment is the antenna. Antennas will be covered in greater depth in the following chapter. However, they are an integral part of the equipment. It allows you to receive signals from the beginning The basic idea is that without antennas is Ham radio.

A basic communications equipment includes headphones, a microphone and, if Morse code is employed the traditional straight key. Nowadays the keyer and paddle can also be utilized for Morse code communications, as it's faster and easier to operate.

To transmit digital data, you'll require a computer in order to understand the signals. In order to do this it is possible to disconnect from the mic (and headphones if required) and then plug in all the devices required.

Filters

A few hams use filters that block certain frequency ranges or frequencies. There are a variety of filters that are based upon what you wish to accomplish.

1.) Feedlines serve to prevent unwanted frequencies from passing through the antenna at either end.

2.) Receiving filters are located into the radio. They typically consist of quartz crystals that allow only the signals that are desired to get through to the receiver.

3.) Audio filters typically assist in removing unwanted background noise

4.) Notch filters can be used to eliminate a small frequency range.

Basic understanding of radio terminology can be acquired by studying the topic and understanding which bands are assigned to amateur, also known as ham, radio users to utilize.

There is no way to simply transmit using any band you want, as many bands are dedicated to particular industries, or to particular purposes like to communicate with government officials or in emergency situations. It was mentioned before that the ham radio is able to be a focal point during emergencies, like in the aftermath of 9/11 or the hurricane Katrina in which amateur radios aid in communication for rescue coordination.

Chapter 9: Advanced Concepts

Once you've got something about the theory of radio transmission, we can explore the deeper ideas. In this chapter, we'll take a look at different types of antennas, signal gains, signal-to-noise ratios as well as licensing requirements for radio amateur operation. Don't be concerned should you find yourself feeling that we're speeding up, just read the chapters that you have to read again, and you can rest certain that we'll reiterate information as we go along while also resummarizing everything at the conclusion of the book.

Although you are just beginning your journey, it's essential to provide a broad review of the subject so that you can comprehend the ham radio in general. Begin small, and then work to advance is the most effective approach but examining more sophisticated ideas will definitely show you the way to progress in the field when time passes.

Antennas

In general, there are three varieties of radio antennas:

- Omnidirectional (also called 'omni')

Axis directional (also known as 'Yagi' antennas)

Panel (also known as flat antennas)

An antenna's construction is constructed in that they direct the transmitted signal to a particular pattern. When it comes to an omnidirectional antenna, the word 'omni"means "all" and the term 'directional' means an aiming of the transmission. Therefore the omnidirectional antenna will aim the signal to every direction.

Though this could be the ideal type of antenna, it is important to remember that since the signal is distributed across many directions, it will lose its strength much faster. Consider it like having the appearance of a tiny pool instead of a river that will be moving in a larger in size and power in general.

The directional antennas look like old-fashioned TV antennas which were mounted on rooftops of homes. They typically comprise of parts made of aluminum, and are sometimes described as fishbone antennas because of the way the antenna components resemble fishbone.

The directional antennas channel the radio signal into one slice-shaped, pie-slice shape. This is essential as, in order to ensure the greatest reception possible for your radio amateur setup, it is essential to be aware of the direction your desired signal is emanating, and then you must position the antenna in that direction. Failure to achieve this, basically is a sign that you're off frequency, not receiving signal, and in fact totally off-kilter. You can do this through obtaining GPS coordinates and entering that data into a map, which can provide azimuthal coordinates to the user.

Panel antennas are not likely intended to be a an integral part of any radio station's equipment because they're usually used for transmitting microwave information with very high speeds. A majority of amateur radio operators count on extremely high-gain directional antennas that have many components.

Signal Gain

Like we said earlier, it is possible to amplify an RF signal to make it simpler for the recipient to access the signal. This is done employing high-gain antennas that are not equipped with amplifiers or through injecting power to the transmission line via the process known as an in-line amplifier for RF.

The method employed to increase the volume of sound, the word used to describe the increase in signal is known as decibels or 'db' in short. There's a chance that you've encountered the term decibel in the past to define the quantity of sound. It's not a accidental. How the volume of sound increases with the amplitude of sound is similar to how the power of RF is regulated when more power is included in the hardware for transmission.

The decibel value is an indication of the signal's power to noise ratio. It's that simple to admit it, but you'll be scratching your head while you read more about the topic. Why? because decibels can also be utilized in various other areas such as engineering or acoustics. applications of decibels isn't always intuitive. For instance, in the area of RF engineering, a shift in power that is a fraction of 10 causes the change to be precisely 10 decibels, that is 10 decibels.

The decibel was initially employed to measure signal strength by Bell Corporation during the advent of the phone. The company needed to know exactly the amount of loss in signal was occurring which is why they chose decibels as an effective device. Nearly every antenna you see when you look for an amateur radio set-up will have an a decibel rating. Keep this in mind when you choose your equipment.

AVOID: Amplifying specific RF frequencies could be unlawful in the state that you reside in. Do the research and find out what frequencies you can use in your particular area as well as what their highest output rating is. Don't operate a radio without an antenna connected to it. It could cause the device to heat up due to its energy not being utilized correctly. Making sure to identify these problems prior to time can prevent massive problems in the future and also ensure that you are able to get the unit out of the ground in beginning.

Signal-to-Noise RatioSignal-to-Noise Rat measure of the intended radio signal in comparison to the noise of radio waves that's everywhere. In the past radio transmissions may come from numerous sources, including distant stars within the galaxy! Due to the amount of radio frequency noise to deal with it is important to be sure that the radio signals will have enough volume to stand out among the background noise.

Think of having a noisy space with lots of individuals, each talking to one another. Imagine that you wish to speak over the noise of the crowd with someone who is just a just a few feet further away. The most likely scenario is to increase your volume in order to be heard, don't you think? It's to separate your voice from that of the crowd to ensure that your companion will be able to recognize you more clearly. Why your friends can clearly hear you is due to an adequate signal-to-noise ratio which is achieved and the communication is now possible with no interference.

Licensing for Amateur Radio Operators

Because of the Communications Act of 1934, every amateur radio station is controlled under the Federal Communications Commission, or FCC in the short. In addition, there are many international agreements currently being implemented - these agreements direct the manner in which amateur radio operators is able to communicate with radio stations in different countries. To operate an amateur radio, you must get a license. Generally speaking, there exist three license classes within the United States. The availability of frequency is restricted for those who have taken the classes that meet a specific class, and the better the grade, the higher possibilities for frequency are made accessible.

However, getting the license isn't the price of a ticket, and it's some learning needed, as a good result on a test is a requirement to get licensing at every stage. Although the FCC decides who can get which license, the examinations themselves are conducted by groups of volunteers made up of fellow amateur radio enthusiast. The exams are conducted under the supervision of groups that are referred to as VECs, or Volunteer Examiner Coordinators, also known as VEC's.

Volunteers administer the tests and submit the test outcomes to the FCC and they issue an applicant with a license. The majority of U.S. amateur radio licenses last for around 10 years prior to renewal. The only restriction that could prevent the holder from getting an license is the case if you are representing an international government.

Structure of Licenses

In December 1999 in 1999, the FCC was able to roll out modifications to the way licenses were allocated. The changes were made in April 2000. quantity of licensed amateur radio broadcasters was cut from six to just three. In February 2007 the FCC did not require applicants to pass the Morse code proficiency section. These changes were implemented by the FCC to improve the efficiency of their licensing process more efficient as well as to bring it in line with the technological time we are living in. While these new adjustments have made it a slightly more difficult process to acquire started, processes needed to obtain more advanced licenses are less.

The three classes of licenses are:

Technician Class- It is a basic license that is available to all newcomers to radio amateurs. The exam that must be passed to obtain this license consists of approximately 35 questions. The majority of them are related to radio theory as well as certain operating procedures that are standard. Technician Class license Technician Class license allows the applicant access to all radio frequency bands for amateurs that exceed 30 MHz. This allows users to connect across all of the United States and some other regions in North America. The license allows for some specific privileges when it comes to the usage of short-wave radio bands. They are able to be utilized for some, but not all, communications methods that are international in nature.

It is the General License Class - This class gives the operator some abilities that aren't covered under those with the Technician Class License. Perhaps the most attractive feature of this class is its ability to gives amateur radio operators an opportunity to connect with foreign countries. To get an overall license, the radio amateur operator first needs to obtain an Technician Class license and the next step is to take another exam of 35 questions. It is the one that the majority of operators are aiming for having, because it provides the highest level of freedom in communication.

It is the Amateur Extra License - This class permits you to operate radio equipment for amateur use across all frequencies, and across all modes and modulations. The exam that is the final one for this class is around 50 marks, with a good score it is necessary along with the previous two licenses. It could be said that this is the "big daddy" of licenses.

If you're still unsure concerning how you can get your own amateur radio licence, please go to www.arrl.org to find out more as well as any current time changes, as they happen. While we're able to provide you with current information as of the date we were the publication of this guide However, you must be aware that the information is constantly changing and so a quick check through the official website is the best method to be sure that you have the latest information.

Chapter 10: Your First Contact

Once you've got an understanding of the fundamental ham concepts and you have (hopefully) been able to successfully complete your registration procedure, it's time to get started working on it.

If you're considering amateur radio, chances are you already have your very first equipment in place and want to join in the excitement. We will reiterate this, if you're getting confused or feel that any aspect of what has been discussed in this article isn't easy to comprehend, you are welcome to revisit the prior pages as frequently you'd like. Knowing the entire spectrum of ham radio provides the most solid foundation you can develop in your personal hobby. Also, we'll summarize the entire chapter in a second time.

There are a variety of ham bands operating around the clock therefore your ears are the most effective tool during the initial phases of the procedure. Listen to different frequencies and observe what's happening, acquainting yourself with the environment you're soon to experience.

It is possible to check a range of frequencies by turning the tuning knob of the radio. The current frequency is displayed on the display or dial. The frequency will depend on the type of signal; the method for tuning may differ. For FM or Morse codes that are transmitted by radio, make use of your ears to hear to the signals, but in the case of signals that are broadcast through specific equipment, you'll require a display in order to determine exactly the right frequency.

10 12-15 17, and 20 meters typically allow excellent QSOs (reported) Long distance (DX) during the day. For the 30-40 to 80, 160 and 30 meters you can also discover excellent DX contacts in the evening which is centered around sunset and sunrise, but throughout the day the number of QSOs can be quite limited.

Common types of signals

Morse code (CW)

Everyone has heard about Morse codes from the past but you may not are aware of it or it is. Morse code, sometimes known as Continuous Wave (CW), can be tuned in using this easy method First, you must set the rig up to receive Morse code (switch the rig to CW mode) Then, if there are additional filters available then set your rig to use a broad filter. Adjust the knob of tuning until you can hear the sound. When you have found the signal that you're interested in then you are able to apply a filter that is narrower to block out all noise to make better listening.

The discussion will focus on Morse code more in the following chapter, providing you with more information and some background to add some knowledge and curiosity.

Single-sideband (SSB)

This is the most popular mode used for voice transmission in radio frequencies. SSB is highly efficient and helps save radio spectrum. If you want to be able to detect the SSB signal, you must first configure your device to tune in the right way. After that, use the largest possible filter. Turn the dial on and off until you come across the frequency that you're looking for.

It is usually possible to tell that you're closing in on the active SSB signal through hearing the sound of crackling or rumbling. At this point keep fine-tuning until your voice is distinct.

Modulation of frequency (FM)

FM has become the well-known way to transmit on both the VHF as well as UHF VHF bands. For tuning into FM, first you have to configure your device for receiving FM signals. You will then need adjust the squelch order to be sure that you be able to hear weak signals however, avoiding interference.

Then you're ready to begin your search. Enter your frequency on the keypad or type it into the dial for tuning. When you find the frequency that is active it will be possible to be able to hear the voice of the operator. It is possible that you will have to go through a few back and forth movements until you reach the frequency at which the voice sounds the most clear. FM radio is renowned for being the ideal place to go for the top radio stations. Usually, they are for news and music.

Radioteletype (RTTY)

RTTY is a particular kind of signal. It is transmitted in the result of two tone marking and space. The signals can't be easily decoded (by the ear) or by ear, so they need an external device that can listen into.

For tuning to RTTY You will initially be required to set the radio to the appropriate mode (RTTY or DATA mode). After that, you will need to configure the decoder/encoder for data. It will be based on the model of equipment that you are using You will likely have to refer to the instruction guideline to ensure that you are doing it correctly.

Once everything is in place when everything is set, you are able to adjust the tuning using the same method as SSB however instead of focusing on the radio, you'll must keep an eye on the indicator for tuning and watch whether the tones are aligned correctly.

Making the connection to the RTTY as well as DATA signal can be a little more complex and will require extra equipment. You are likely to want to hold for a while before trying this until you've gained more experience. Basically it's more than hearing and has more to do with the technique.

HF, VHF, and UHF

The high frequency (HF) bands are categorized into two main segments. The lower portion is typically used for CW as well as information (including RTTY) transmissions while the upper end of the frequency can be occupied by voice communications. If you're looking for distant contacts, you'll most likely encounter these in the lower portion. Casual conversations, also known by the term "ragchews," are found on the top portion of the segment.

Contacts for VHF as well as UHF channels are made with the aid of repeaters. They are organized into series of channels. A majority of the ham operators using these frequencies utilize FM channel (frequency modulation). It's a fantastic option for vocal signals as it helps to reduce noise and, consequently, provides a an enjoyable listening experience.

With repeaters, ham radio operators are able to travel long distances using their messages. Most often, the communications are of a more intimate nature as opposed to those that are found on HF. They can be used to keep connected to family members and close friends. For the purpose of making use of a repeater it is necessary to turn on the access to tone in your device. This way is to let it know your signal was intended to it. If not it won't send the signal any further.

If you're looking to test and get the repeater that is in your region then you must comply with these instructions:

1.) Locate a local repeater near the area you live in. It is possible to do this through an online directory or an easy search on the internet that will provide you with a details.

2.) Find out what frequency is for both incoming as well as outgoing signal.

3.) Set your radio to the correct frequency and set it so that it's listening on the frequency of the repeater's broadcast.

4.) Tune in the same manner just like you would normal FM signals, and continue until you get a real signal.

Types of contact

There are three main types of contact (QSOs) you'll come across when operating an amateur radio And we've already discussed casual chats ("ragchews") nets, and contesting.

To recap the fact that, being an amateur radio operator, you'll communicate with others across a range of frequencies, or bands even those that specifically designed for use by ham radio. They are identified in terms of wavelengths, with one of the most popular one being HM (High Frequency) or short waves. If you want to contact someone else, i.e. an amateur radio operator the first step is an QSO. A QSO is an integral part of what is known as the Q code. It is a global deal and can be used to Morse code. This is something that we we will be discussed in greater detail in the following chapter.

The QSO acts as an introduction. It's almost similar to telling someone "I'm here, does anyone want to speak?" This is a great approach to speak to individuals who don't know your native language because it is a common spoken language within the sport, and usually created by an amateur looking to talk to anyone new or someone they had "met" before.

When another participant hears this phone call and wishes to talk, an exchange of information begins, followed by it will be decided on whether the contact was actually established. If that happens it is possible to go into conversational manner, exchange QSL cards, and generally create the acquaintance of a new person.

It is not surprising that conversational conversations are the much of the conversations you come across as you traverse diverse frequency. The exact origin of the phrase "ragchew" (i.e. chewing on the of the rag) is not clear but it is an accepted term for ham operators.

If you do decide to venture out into the wilderness (which will be described further down) Ragchews can provide a good learning ground. They're informal and allow you to discuss any topic you want to discuss. In addition, they can assist you in gaining confidence and skills that you require to grow the ham radio skills of your future.

The nets (networks) are meetings between Ham operators, which take place consistently with pre-arranged dates. The nets are organized around a particular area or a set of topics they cover, with numerous discussions on the operation of a rig, as well as the technical aspects and related issues. In essence, it's an excellent way to gain more about the subject, increase your confidence as well as get important tips and suggestions from more knowledgeable Ham radio operators.

If you've found the right website that you're keen on, you'll be required to comply with certain rules for participation. The first step is to check to your NCS (net Control Station) who determines the time for meetings and durations, as well as other rules. If you're curious about what the internet is for you as a user take a look at what their policies regarding this subject are. Also, find out when they are open to visitors, and what's to be expected from you. Be sure to adhere to these guidelines, since this is the standard radio manner of conduct.

Contesting can be a really interesting method of communicating. There are various competitions where hams attempt to swap short messages as well as their call signals as swiftly as they are able to. Similar to the pursuit of DX is an additional type of competition in which the contestants attempt to contact remote stations and operators.

These are the basic rules to be followed when participating in competitions. Once they must be observed to prevent exclusion, while also keeping your image good.

1.) Keep your communication brief. It is important to create as many contacts as you can within the least amount of time Therefore, you must be sure to keep your communications short for this purpose.

2.) If you are DXing, swap only the information that is required before moving on for the station next.

3.) Prior to participating in an event, make sure you make sure you check out a certain station to find out what the contest's focus is, i.e. what kind of data they normally share. Sometime, it's your exact location as well as the signal report. However, it may be a serial code, and so on.

4.) The communication you are having with individuals from around the globe. This means that you will come across people who don't know English. Most of them will be able to communicate some basic information. However, in the event that they don't have the necessary knowledge, they can utilize the Q-signal set that is international.

Q signals

Though most people can know at least a little English in the present, understanding the basics of Q-signals could be extremely helpful as a last option. These abbreviations are derived from the very beginning days of radio. Any amateur worth his to know should have a basic understanding of these. Signals of Q are commonly used in conversations between English users as well for making transmissions more rapid.

Below, you'll find some of them as you to refer to, in order to increase your interest and provide some basics for starting. An entire set of Q-signals may be quickly found online and with no effort. These are easily learned by observing.

QRG: asking about the frequency of your device?

QRN: Are you getting static?

QRT: Should I stop the sending?

Are you prepared for QRV?

QRU - no issues to discuss

QRZ - Who's calling?

QTH: What's the location of your QTH?

QRX - We will call to you once more

QSB Signal Fading

www.ingramcontent.com/pod-product-compliance
Lightning Source LLC
Chambersburg PA
CBHW071439080526
44587CB00014B/1915